The Definitive Guide to Green and Organic Pest Control

The Definitive Guide to Green and Organic Pest Control

By Andrew Dzieman

No part of this book may be reproduced by any mechanical, photographic, or electronic process, or in the form of a phonographic recording, nor may it be stored in a retrieval system, transmitted, used as part of, or whole of an internet site, uploaded to the internet, converted to a digital format, or otherwise copied for public or private use, without written permission from the author.

Acknowledgements and Dedications

In my first book I had the joy of thanking some very specific people who helped set the example and those of which I base so much of what I do. But the last year of my life since I published my first book has had its up and downs. I need to thank everyone who motivated and pushed me forward, allowing this book to be finished. The largest motivational moments in the last few years have been the phone conversations with my brother Charles, my mother Margaret, and my best friend Joe Ybarra. These conversations about life have led to more understanding, and wisdom for everyone involved. I would also like to thank Oscar Campos for helping with the proofing of this book and getting me to the final stage of publishing a finished product as opposed to something half-done.

The Church at Rocky Peak has shown me the rewards of faith and the responsibilities associated with real spirituality. Joining the family of Christ followers there is one of the greatest decisions I've ever made.

All of my friends in the wrestling business, especially those from Pro Wrestling Guerrilla, and the Empire Wrestling Federation have reminded me of how often hard work and dedication leads to success. "A good carpenter never leaves home without his tools" is the hallmark statement at the School of Hard Knocks. Jesse Hernandez sets the example with the right kind of energy, unwavering tenacity, and a type of wisdom setting him apart from everyone else.

The last area of personal motivation has come from my time at the XplosiveFit dojo. I need to give a special thanks to all of my fit family and to Coach Rod and Vanessa for leading so naturally. To Shane and Aileen for providing much needed support, and to Ed Peisner, who's motivation through our workouts has been tremendous while his work with the Jordan Strong Foundation is a constant reminder of how important a happy life is. Everyone at XplosiveFit has been gracious and there is no shortage of positive energy at the dojo.

I would be nowhere without my wife and two daughters. Laura, Zoey and Madison are the greatest motivators and they constantly drive me to do more and be more. Lastly, I need to thank the father of creation, the son of God, and the holy Spirit. I have been blessed not just to live on his earth, but to be a part of the American experiment, and for that I am truly thankful.

Disclaimer Clause
or
Statement of Warranty

The author and anyone associated with the writing or publishing of this book are in no way responsible for the application or use of the chemicals mentioned or described within. Likewise, the use of all chemical products, non-chemical products, and any listed items are done with the understanding that the manufacturer determines the proper uses, and when conflicts arise between this book and the manufacturers recommendations, the manufacturer's directions should be followed, and this book's instructions should be ignored or modified to match the product manufacturers guidelines. Those associated with the production of this book make no warranties, expressed or implied as to the adequacy of any of the information presented in this book, nor do they guarantee the current status of registered uses of any of the chemicals with the U.S. Environmental Protection Agency. Also, by the omission, either unintentionally, or from lack of space, of certain trade names and of some of the formulated products available to the public, the author is not endorsing those companies whose brand names or products are listed. Product names may be copyrighted and in all cases the product names are the property of the manufacturers of those products, or other respective owners.

Contents:

Preface

Chapter 1. Defining Green Pest Control

Chapter 2. Integrated Pest Management within Organic Pest Control

Chapter 3. Planning Services

Chapter 4. Non-Chemical Control Methods

Chapter 5. Pesticide Applications within an organic pest management program.

Chapter 6. Borates for Wood Preservation and Control of Wood Destroying Organisms.

Chapter 7. Strategies and Case Studies

Chapter 8. Conclusion

Preface

So many people are looking for "simple" solutions to pest related issues and if you search hard enough you what you will find is there are in fact, a lot of "simple" solutions. The downside to this availability of "simple solutions is the average person's lack of understanding about how to control pests in the most environmentally conscious manner. Product labels often go ignored by people with little understanding of how products work or how pre-designed processes should be put forward. In fact, I see the results of these "simple" solutions all too often. "Results may vary" should be printed in large text on the cans, bottles, and jars of pest related products which are so freely available, and it should be printed within every internet article promising the next big thing in pest control.

Pest control should be a process by which pests are at least mitigated, and benefits should always outweigh risks. The point of this book is to take the risk versus benefit idea a step further. Risks should always be minimized. Therefore, careless over-application of products should be eliminated, and the primary focus should be on as much control as possible through non-chemical procedures from the very beginning.

Our world has moved past the carelessness of the 1950's which led to the banning of DDT. Environmentalism has become entrenched in our society, and everyone participates in it, even if they don't believe in it. Sustainability, emissions control, and chemical awareness are just some of the new ideas affecting our behavior every day.

While working in the pest control industry I've put a lot of time and effort into figuring out what organic pest control actually is and how it's applied. I learned one key thing an entire industry of technicians and salesman seemed to miss; how to properly apply organic pest control. After training through Purdue University and becoming certified through the Entomological Society of America, I decided it was time for somebody who actually understands it makes the effort to explain it.

One thing you will notice in the book is that I make a conscious effort to move past using two specific words. "Green" and "control" are both problematic words. Green is in fact a color and not much else. There are no official standards for green pest control and therefore there is very little for me to write about regarding green pest control. As for the word control, I've simply replaced it with management. Organic pest management is a more appropriate term which reinforces the inclusion of those principals belonging to integrated pest management. The inclusion of IPM principals in organic pest control are necessary components which help us achieve a better result with a lower impact.

This book is my attempt to put forward all of the strategies of organic pest management and provide a training guide to those who need one, while showing non-trained individuals what they should be looking for in an organic pest management program. The knowledge contained in this book will provide the average person and seasoned pest control veteran with everything they need to judge the value of the organic pest management work they over-see. I hope this book educates those who otherwise wouldn't know about the topics discussed, but I do also hope it provides an avenue to better, less toxic pest control methods as a form of common practice.

Chapter 1 Defining Green Pest Control

The Idea of Green Pest Control

Pest control, which can be broadly defined as the eradication or mitigation of pests incorporates many techniques. The goal of any pest control program is to lower infestation levels to specific numbers pre-determined and established by the scenario and environment of those affected by pests. Pest control however, has been widely criticized for the incorporation of toxic pesticides and their over-use. The irresponsible manner in which pesticides are used as the primary control method takes into little account the inherent value of a job done right. In many cases control has been achieved with almost no thought about the side-effects resulting from the haphazard way pesticides are applied. The use of pesticides as a broadcast treatment covering large areas makes almost no assumption about the negative effects possible to occur and applying large amounts of pesticide into more confined areas can cause more harm than good. The average homeowner, and even the average professional applicator can cause unknown levels of harm, only becoming known after several years in the cases where the harm is discovered at all.

Green pest control is thought of by many as an environmentally conscious solution providing an alternative to traditional pest control treatments and the word green is used to imply it is a more environmentally sound solution. Green pest control when put to action however, does not have a unifying set of standards. In fact, green pest control does not have the same definition to everyone. "Green" can in many cases be used as a marketing term by those looking to sell their products and services, or it can mean to some the exact product of organic pest control. Whichever way the term green is interpreted should be suspect when used for marketing or used in a vague manner. Organic on the other hand, is a specific term. Its primary use is almost always to describe a specific process, and commodities created under the process. The term organic does however mean very little with regards to safety when describing pesticide products. Organic, therefore can have two distinct meanings. As a process, such as with organic pest control, it is meant to imply a certain level of safety, especially when compared to non-organic methods. As for the chemicals used in the process, the term organic does not imply safety.

An organic pesticide in some cases may be just as dangerous as a synthetic one. Therefore, further research beyond a pesticides composition may be needed to determine a pesticide's safety level.

The pesticides used in organic pest control are often referred to as biorational pesticides. The term biorational implies products or active ingredients with a lower level of environmental impact and a relative expectation of higher safety. Biorational pesticides tend to have very low risk factors to the point where the benefits of treatments outweigh the risks associated with their use and then provide for extra levels of safety. Organic pesticides are related to organic pest control but are not a necessary part of it. Even though many organic pesticides, such as plant oils are considered biorational, organic pest control can be accomplished without the use of pesticides. Because of the many steps required and the prerequisites of organic pest control, using an organic pesticide alone does not equate to performing organic pest control and if one of the following terms, organic (as a strategy or process), biorational (with regards to products used), or IPM (integrated pest management) do not also apply to the treatment, then most likely the treatment should not be considered green or organic. All 3 terms are intricately connected to organic pest control and thus the most liberal definition of green pest control should also include all 3 because all 3 are strictly required for proper organic pest control.

Connecting Green Pest Control and IPM

A green pest control program should be as intricate as possible. Planning should be heavily focused on the details as well as the larger picture. Focusing on both the micro and the macro provides for almost every necessity, prepares for almost every mistake, and considers the environmental impact of every aspect of the service. Because all pest control services have some form of environmental impact, care should be taken to determine environmental issues which may occur from treatments. Many environmental factors are obvious and others aren't often thought of. Proper care should be taken to account for even the most unlikely issue.

The impact of pesticides within an intended treatment area can be obvious. There can be many wanted and un-wanted effects in the treatment area and environment including some effects on non-target pests. Many non-target plants, animals, or insects can be hurt or killed as a side effect of treatments and these non-target organisms might be beneficial to the area, or extended environment. Special care should be taken because drift and run-off pose a major threat, and cause damage of which pesticide applicators never become fully aware. Drift and run-off are those situations where wind or moving water take pesticides to areas of which they are not meant to be. These areas can include neighboring properties, and off label application sites. The existence of these hazards, despite the large amounts of work to prevent them, reinforces the basic, unifying standard of every green pest control program; an intentional effort to limit the use of pesticides. In the cases where pesticide applications do become a necessity, pesticide selection should be focused on low toxicity products, which have little to no environmental impact, and control pests at reasonable levels when applied properly. In every case, non-chemical methods should be exhausted or proven inadequate before pesticides are used, and when pesticides are used, only those deemed biorational should be considered. This approach which asserts firmly the concept of pesticides used as a last resort is called integrated pest management or simply IPM. When the inclusion of biorational pesticides as a last resort is added to the process, the process then becomes organic pest management.

Integrated pest management was developed to be a thorough, precise, environmentally conscious approach to pest control in which multiple factors are considered with a goal of limiting the environmental impact associated with pesticide applications. This process provides a steady, thorough, environmentally conscious path towards any pest related goal and is an inherent part of any organic pest control program. It provides standards for almost every part of an organic pest management plan. Organic pest management adds to integrated pest management by requiring more specific pesticide selections through the inclusion of biorational pesticides and mandating additional steps of care to be taken with pesticide use.

Before integrated pest management and organic pest management were common, specific goals usually went undefined and treatments frequently occurred in a sloppy haphazard way. With integrated pest management, and organic pest management however, goals are always defined, and treatments are more precise. Many would consider Integrated Pest Management to be the cornerstone of organic pest control. Any green pest control program should contain the standards of integrated pest management if it is to be taken seriously as a more environmentally sound solution. Non-chemical methods such as environmental alterations are to be made with the goal of either excluding pests or making problem areas undesirable to them. When pesticides are used, product selections should focus first on the environmental impact of its use and human or animal health concerns. Integrated Pest Management alone has no standard with regards to the use of organic pesticides as part of the IPM Plan. This is why integrated pest management as a complete process with generic pesticide application is not considered a green or organic pest control solution. On the other hand, IPM with no pesticide use is virtually always organic, safe, low toxicity, and complete. Therefore, pesticide free integrated pest management programs can be considered green.

Environmental considerations

The benefits of pesticide applications should always outweigh the risks, and environmental damage should be avoided at every step. With professional pest control work for hire, customers will sometimes have specific requests such as organic and biorational pesticides or in many cases, they may request "the strongest" pesticides, as they would define them. Many homeowners treating houses themselves will also have specific products they prefer over others. Anyone who can't determine the safety of a given product should not use it and pest control technicians should be able to use their own best judgement. All attempts should be made to avoid over applications and poor product selection when pressed by customers and those less knowledgeable. In some cases, customers will request pesticide mis-applications assuming the result of higher product amounts, or off-label site applications will have greater results. Technicians and professional pesticide applicators should communicate the safe and effective approach of IPM to their clients, and those otherwise affected. It is important for those not applying the products to understand the reason for the chemical applications, the locations, product selections, and how the applicator made his decision.

Integrated pest management is defined by specific standards and they must be followed. These standards are the cornerstone of IPM and are incorporated into organic pest management. Because of this, a more efficient, more effective, and environmentally conscious solution for any type of pest. Organic pest control, as it was once called is now referred to as organic pest management and has become more environmentally friendly with the inherent pesticide limitations. In many cases, control can be achieved without the use of pesticides and the integrated pest management approach assumes at least some non-pesticidal treatment options are performed before pesticides are applied. In the inspection phase of the process many conducive conditions and contributing factors are found and when corrected pest issues may be resolved. Also, proper identification of an insect leads to a higher level of control. Through inspections and research, the reason is often determined for the presence of a pest in a specific location and what changes can be made to correct the issue most reasonably. Likewise, previously undiscovered, ongoing issues may be unearthed through a good inspection where necessary corrective measures such as fixing leaking pipes, filling cracks, replacing damaged wood members and others are discovered.

As part of the process, those affected by insects are expected to define their tolerance levels. A tolerance level is the amount of pest infestation occupants on a property can withstand before pests become a nuisance. The point at which action is taken is called an action threshold. At this point, pests are no longer within tolerance levels, and treatment usually becomes appropriate. In many cases, there may be a specific action threshold given before pesticides are applied. If infestations persist beyond a certain point in terms of pest population, or time of infestation then pesticide application may be deemed the appropriate action. Reaching an action threshold may not always imply a specific action, it merely dictates one should be taken and the strategies relied on within IPM allow pesticide applications only when appropriate. This includes reaching a pre-determined threshold level. Pesticide applications are therefore a last resort or are deemed necessary in cases where other measures won't be or haven't been reasonably successful.

Chapter 2 Integrated Pest Management Within an Organic Pest Control Program

The Use of Chemical and Non-Chemical Methods in Green Pest Control and IPM

IPM (Integrated Pest Management) is a solution incorporating multiple techniques for enhanced control with a lower impact on the human health and the environment. Due to the multiple steps involved in the process chemical application is less important and monitoring the property for pest activity is necessary. Pesticide applications, when they occur should be as precise as possible and always have a specific reason. Large broadcast treatments and precautionary over-treating are considered irresponsible due to the possible transfer of pesticides to non-treatment areas and they violate the self-imposed limits which are an integral part of IPM. Crack and crevice treatments, spot, and void treatments are considered a better solution and have replaced the "more the better" approach to pest control. Crack and crevice treatments tend to be the most effective and are done using lower amounts of pesticides. Since a product can be applied directly under baseboards, and deep into cracks it limits the exposure to the applicator and the occupants while enhancing effectiveness. Void treatments which are also considered low impact are similar because there is little exposure to the applicator and those within a structure because the treatment is very targeted.

Integrated Pest Management relies heavily on planning ahead and following a set of steps including 1. Inspecting and monitoring infestation levels 2. Assessing the problem and determining the level of infestation 3. Determining tolerance levels and setting action thresholds 4. Implementing the pest management plan which starts first with non-chemical control methods and may continue with pesticide applications 5. Evaluating, monitoring, and communicating to assess the success of the pest management plan.

Inspecting and determining the problem are the most crucial steps in the process. If an insect is misidentified, it's reason for being there and the proper method for control may be lost to the individual tasked with performing treatments. Inspecting properties should always be done before any treatment strategy is put into place. This starts by walking the property and documenting every pest problem and related issue. The pest related issues which may lead to an infestation or may be the cause of an infestation are referred to as conducive conditions. Identifying these conducive conditions can in many cases be more important than the actual evidence of insects. Corrective measures for insect related property issues can control insects by limiting their access to a food supply, a water supply, and a harborage source. Likewise, correcting many structural issues will change insect behavior in a particular manner lowering populations in concern areas. If fewer insects can enter a structure, then it's likely less of the common invaders from outside will be noted inside.

Establishing Tolerance Levels

Pest tolerance levels are the levels at which a pest problem becomes intolerable and specific actions are warranted. As an example, one ant may not pose any worry or threat within a structure, but a trail or several trails could be the point at which pesticides are used during a pest management plan. In many cases a specific trigger may lead to pesticide application because of a pre-determined action plan. This is referred to as an action threshold level. Implementing a plan is the most important step in control and pesticides are not always the best solution. In some cases, an action threshold level can trigger a specific non-chemical approach such as cleaning, or re-inspecting, and if pesticide application is then warranted based on tolerance, and other methods exhausted, pesticides can be applied.

Some people will have their own physical and mental aversion to pesticides. This aversion should always be a consideration and pesticides should never be used around chemophobic persons and those suffering from real chemical sensitivities. Chemical applications within an organic pest management plan are further restricted to limit and sometimes prevent pesticide exposure to an additional degree when compared to standard forms of pest control. IPM and organic pest management allow those performing treatments to be creative with non-chemical measures. One example is with rodents entering a structure. Rodent access can be limited in many cases with repairs of structural defects preventing their continued entry. If tolerance levels are set concerning the rodent population within the structure, then a lack a of rodents means no further treatment is necessary.

Looking for and documenting the results of treatments allow for a better understanding of control levels on the property and of further issues. There are several methods of monitoring results including but not limited to inspections, examinations of monitoring devices and interviewing those affected. Pest monitoring and evaluating treatment results. This allows for better control by providing consistent knowledge of the pest situations on a property. If a treatment isn't working and a decision to change methods is required, the change may be better facilitated if done as soon possible. Without knowledge of treatment failures, or shortcomings however, necessary changes in strategy can go undone for indefinite periods of time. Logs of pests seen on a property, and monitoring devices such as glue traps are recommended if not mandatory in almost every pest situation for this reason.

Pest Monitoring

There are several types of traps used for control and monitoring of pests. The most common are glue board style traps with a glue-like substance on a flat surface designed to trap insects, spiders and rodents in some cases. These are sometimes open with no protective cover while other styles fold closed with a more complicated design shielding the glue from contamination. These traps have a very limited lifespan since they tend to fill up quickly with insects, or spiders, and collect dust from the environment around them. These traps can sometimes come pre-baited which means pests are attracted to them with no further alterations. In some cases, an attractant is required. Many common foods work as an attractant for varying types of pests, such as peanuts, or in some cases, cheese. Another option for many types of insects is the use of pheromone attractants. There are 2 general types of attractants. The first of which is a sex pheromone used to capture males of a species by luring them in as though a mate were nearby. This is the most common type used commercially but in the case of residential pest control, and food storage it is most often used for moths. The second type which is an aggregation pheromone is most often used on beetles. Beetles, such as carpet beetles tend to move to areas an aggregation pheromone is present. In nature, other beetles of the same species will release this pheromone attracting each other.

Insect light traps, or ILT's are used at nighttime, inside, or in dark areas to attract insects. They have a bright light source designed to bring in flying insects by using the insect's natural attraction to light against them. In some cases, an electrical charge will hit insects as they approach the trap, and in other cases a glue board is placed near the light source to catch them. The key to successful control with ILT's is proper placement. Traps placed inside opposite a door may attract insects inside. Traps placed outside may not cover a large enough area to control the problem. As part of an IPM plan traps should be placed where appropriate, and in large enough numbers to be effective. In some cases, one trap is enough, and in others more may be warranted.

As a property is treated and maintained, pest sighting logs, monitoring devices, and documents providing for a consistent recording of issues become important. What is done with the information contained within this documentation is also important. Communication between those servicing and those on serviced properties should be consistent. If what is being done and what must be done are not communicated properly to all those involved, the treatments may never be successful. Open communication is the key to success and communicating not just what each person's responsibilities are but what each person can expect from the service prevents disappointments and misunderstandings. Those on a property also need to be educated on the service being provided to them, and they need to be educated about the properties they own or of which they are left in charge. The issues on a property are most likely unknown to homeowners and property managers, or most likely these issues have not been thought of a problem. Therefore, peaking with the customer regularly will allow better correction of conducive conditions and better control on the property. Communicating both verbally and in written form is important. Communication starts with communicating what the plan is, what the results should be and with whom each responsibility belongs.

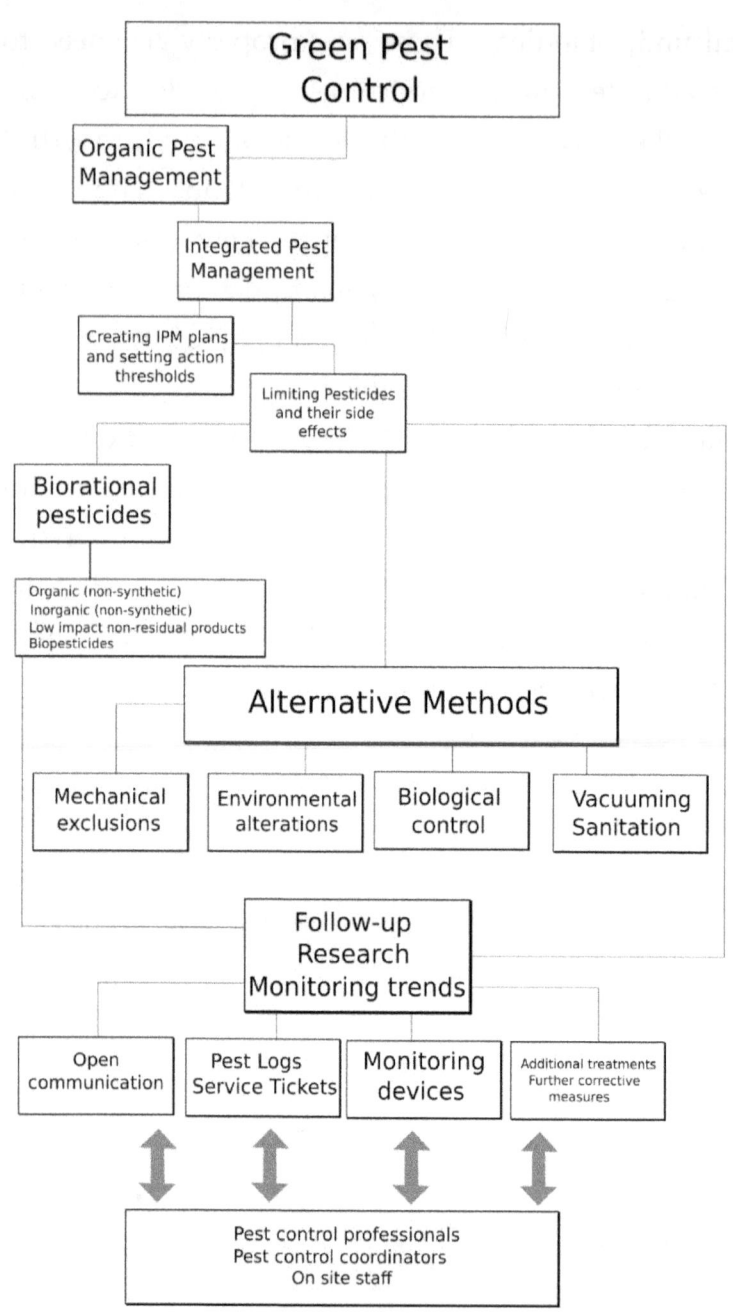

Chapter 3 Planning Services

Professional Technicians and IPM Plans

A detailed planning of services is the best way to assure the desired level of control is reached on a property. When performing IPM based services like organic pest management, one of the most important variables in planning is the determination of whether and when pesticides should be used. Because action thresholds are pre-defined, pesticide applications always have a defined purpose.

Setting action thresholds may be as simple as defining specific numbers of insects found on a property before pesticide applications become appropriate, but also may be more complicated by defining which pesticides are used at which levels and may include an early stage where monitoring begins after pests are first discovered. There can at times be several action thresholds based on different criteria and for different actions. This planning adds to the organic pest management program by preventing unnecessary pesticide applications, forcing inclusion of strategic non-chemical treatments, and providing for higher levels of control.

A professional technician will often devise a series of IPM plans which can be used for different types of treatments. For complicated accounts a more specific IPM plan should be used and written specifically for each property. An IPM plan may be as simple as the instructions listed on a professional pest management agreement and any special notes listed within. It may also be a complicated written report including a site map with specific instructions listed by area. The most diligent pest control companies document every finding and recommendation on a property, save those findings within customer's files, and take pictures of the relevant findings and kept for use in formulating specific and customized treatment plans.

IPM plans can be devised and used by whomever is performing services and can be made in conjunction with on-property IPM coordinators. A well written IPM plan should define the specifics of treatments to be done on the property, and it should note locations of bait stations, traps, glue boards, and other items used for monitoring. An IPM plan should always be used for commercial properties and can be customized for use with residential properties. For commercial accounts a current, accurate site map should be used and adjusted as needed. The plan should always be updated when changes occur to the property or when changes occur with regards to the service being performed. Finally, the plan should define which responsibilities belong to the pest control technician and which belong to the customer.

IPM Plan Specifications for Homeowners and Professionals

Homeowners should pay attention to pest related issues on their properties, focusing on conditions which could add to the severity and likelihood of an infestation. Their own homemade IPM plans should focus on correcting all possible issues before applying any pesticides and as part of a professionally drafted plan, corrections may be required before any pesticide application takes place. Anyone performing treatments can make issues worse by not correcting problem areas and then applying pesticides. Many pest problems can be solved or lowered by correcting known issues however, applying chemicals without correction can often lead to failed treatments. Correcting known issues is an often-ignored part of the process with pesticide applications being performed so easily, but in many cases correcting these issues can lead to more control than applying pesticides alone would ever accomplish. Within a properly applied IPM plan the application of pesticides is less of a necessity and becomes a need only when all other options have been exhausted. Homeowners, occupants, and property managers should remember to define their own pesticide tolerances, and pest activity tolerance levels, verify the correction of all issues, and monitor the levels of pest activity for changes regardless of who performs the pesticide applications.

Liquid applications of products can be either a huge benefit or a massive problem. Many pesticides have repellency levels. Repellent products are the most troublesome because they can cause unexpected, and often unwanted results. Repellent pesticides repel pests as the name implies. In the case of insecticides, repellent pesticides cause what's referred to as a flushing action. This flushing action drives insects out of hiding in many cases but can also drive insects deeper into hiding. In the case of cockroaches, the application of repellents can move these highly adaptable insects from room to room making a small, localized problem eventually fill an entire house. Some pesticides on the other hand have no repellency and some types of products, such as baits are meant to attract pests. It is important to understand the repellency levels of products applied, and to know what the results will be of an application.

When treating for cockroaches, treatments should focus on sanitation, environmental modifications, the application of baits, the application of dry dust products, and selective applications of liquid products. Flushing agents are used by some pest control companies but have been linked to treatment failures when overused. Treatments for cockroaches are complex and after cleaning and vacuuming, flushing agents can be used in small amounts to find nesting areas otherwise hidden. Flushing agents alone are not a complete solution, and when used in conjunction with baits can lower the effectiveness of bait products because cockroaches lose their appetites, causing them to eat less of the insecticide baits. They also move deeper into walls away from the products applied and reproduce more heavily according to some studies. These side effects of repellent pesticide application should discourage the use of these products. Repellent pesticides are used heavily for their contact kill effect. They generally leave little to no residual and if they do, their residual would most likely be repellent as well. One of the main characteristics of organic liquid sprays made from botanical and essential oils is unfortunately their high repellency. This is one of the reasons for non-chemical methods being performed first. Resulting from the complex nature of pesticides and repellency, product selection must be specific to each job.

Repellent products can also have a detrimental effect on ants. Ants along structures can be forced inside with even the minor use of repellent pesticide products. An otherwise small problem of an infestation is enhanced as a result and new problems are created. Likewise, repellent products used inside can drive ants into new areas of the home and divide colonies. Divided colonies will then grow and eventually two colonies can exist the size of the first one. Treatment at this point becomes more problematic. Those species of ants found nesting inside of walls may have multiple queens which occurs frequently with pharaoh and odorous house ant infestations. These budding species thrive while moving from room to room, reproducing and creating new colonies. Exterior nesting ant species such as argentine ants and fire ants get trapped inside trying to avoid repellent pesticides and water issues. In some cases, trails can be seen inside for up to 10 days or more because of repellent chemical applications. IPM standards such as controlling environmental conditions, cleaning, and limiting pesticide use are important in every pest control situation and ants are great example of what can happen when things are not done properly.

Communication

Open communication is expected throughout any pest management program. An integral part of the program should be spoken and written communication between those performing services and others involved in control. Responsibilities should always be pre-determined and appropriately assigned. Many homeowners have their own handymen, and contractors who perform work for them. In many cases a homeowner will do some work himself prescribed by a professional pest technician, or have it done by others. If there is a service technician, he should always verify the quality of work done, and make new recommendations when appropriate. With issues of sanitation it will most likely be the homeowner, or occupants with the responsibility of cleaning except in the few cases it falls into the jurisdiction of a pest control company. This is generally the case with rodent urine and feces which may require a professional program.

For contracted services, such as regularly scheduled pest control, communication forms may be left on the property to communicate problems, document issues, and verify their correction. The first type of form is pest problem areas log. This form is used by those on properties like hotels, restaurants, and other commercial, and semi-commercial settings for employees, tenants, and management to document what they have seen, heard, and otherwise found. There is usually a place on the form where pest technicians can verify the corrections of issues and what measures were employed to correct it. This same form can be used for those on a property to state what they did to correct issues they found and communicate to the professional what steps may still be left.

Trend logs are another form used and should be completed on a regular schedule. Trend logs are filled in by pest professionals, or coordinators working on the property. They are used to document pest trends for the sake of maintaining control and discovering potential issues before they become out of hand. If pests are treated in the same area on a regular basis then conducive conditions may exist, additional or differing treatments may be necessary, or other research into the issue may be required to solve the problem. Without following trends, it many issues will missed. These two forms can be used by pest professionals, but the same forms can be used in many other cases. The trend log can apply to any property, even in cases where homeowners or tenants do all the work themselves.

Professional pest control technicians use service tickets to communicate, and document which products were used on the property. While non-professionals generally aren't required to use these tickets, it's a good idea to document which products were used, how much, and at what percentage of active ingredient in every case. This can be beneficial later in determining why a treatment strategy isn't working. Too much pesticide, or too little can cause resistance, and other I'll effects. If issues occur with one specific product, the another should be used. Trend reports can be used for comparison, and when compared to service tickets, should show trends lessening, and insect populations lowering. If the opposite is true, there may need to be adjustments to the treatment strategy.

For many pest control accounts, pest control technicians may create a site map with locations of ongoing issues, monitoring devices, and other items requiring future thought. Site maps should be maintained with new items added and unneeded items removed. This site map is also a good idea for most other types of properties and can be used as a report of current and ongoing issues and should be checked and updated whenever necessary. A site map is appropriate for almost any property and does not only need to be made by pest control professionals. Homeowners, and other do it yourself technicians can use it to note important IPM related issues, and as part of their ongoing IPM plan.

Chapter 4 Non-Chemical Control Methods

The key to controlling pests in a progressive manner is intuition. Anyone servicing a property, whether the homeowner or a professional technician should know which insects are common in which season, or even which months of year, and what route to take in controlling them based on the situation on the given day of the year. A progressive service is done monthly, bi-monthly, quarterly, or whichever frequency is most appropriate for the property being treated. Professional technicians should know where to look when inspecting, be able to devise standards, and create a plan to correct pest issues. Homeowners should research their own properties, neighborhoods, and extended area for pest trends, and common issues in their environment. Intuition can be a very beneficial component of any pest plan. This intuition, and research is applied to the service by changing control measures where appropriate. Treatment methods alternate as a result, and as the pest problems change throughout the year, new data is applied to the IPM plan making treatments more effective. This may lead to a scenario where pesticides are used on one service, but not on the following service, or where pesticides, and chemical families are changed throughout the year.

In many common areas of a home, office, or other structure, small debris or water can accumulate providing sustenance for insects, and other pests. Food crumbs on cupboard shelves, and spilled food elsewhere can compete with bait making it less likely to be ingested by insects. Areas with mold, mildew, and other moisture related issues can provide food for insects as well, and a viable breeding source. As part of the very first treatment and many subsequent ones, removing insect bodies, food sources, and harborage through vacuuming will aid in long term control. Also, insect eggs are pulled into the vacuum cleaner and removed from the treatment area when the vacuum is emptied into the trash or removed from the property as a vacuum is taken away by pest technicians. If a homeowner is performing services, or cleaning as part of a pest management plan on their own, the vacuum should be taken outside and emptied directly into the trash immediately. Vacuuming also aids in control when chemical treatments are determined to be a necessity. Removal of insect eggs and dead bodies will help to control the population much quicker, while removing insect food will cause insects to forage and eat the bait products used. Bait competition is common occurrence with many insect pests including cockroaches and ants who may be distracted from bait products by the readily available food sources available to them.

Conducive Conditions

While points of ingress, and insect biology are important in deciding how to control insects, it is also important to look for and to control conducive conditions. As part of your inspection, when you arrive on a property you should be looking for signs of improperly watered lawns, sprinklers hitting the house, organic debris such as leaves, major sanitation issues, water leaks, and other items with the potential to be a problem, even if no pest are visible now. Conducive conditions are those pre-existing issues on a property believed to create or amplify pest problems and should be corrected whenever possible. Dripping faucets, standing water, improperly watered lawns, and other water related issues provide some insects with water but may also provide a food source through mildew and fungus. Many flies and other filth feeding insects prefer decaying organic matter. accumulations of decaying fruit, dead leaves, and even moss on the ground can provide them with a food source.

Open doors, and windows with no screens will provide insects with an obvious entrance to the structure, and a not so obvious entrance comes from doors and windows improperly sealed. Inadequate doors and windows should be corrected with weather-stripping, door thresholds, door sweeps or other forms of exclusion to detour entry. Pipes and conduits going through the stucco are often great entry points for insects, and they should be sealed using caulking or sealants. Damaged vent screens should be repaired or replaced.

Moisture levels on a property are important in controlling pests. Many insect problems can be traced back to too much, or not enough water to keep insects thriving outside. Inside, water related issues cause insects to hide, breed, or migrate throughout a structure. In the summer time ants may come from outside as a lawn is not holding moisture the same as it did in the spring. Ants will also congregate around leaking sprinkler heads outside and will move towards the foundation of a house if moisture is likely to be there. This happens because of leaking hose faucets, dripping air conditioner condensation lines, and other small but significant moisture related issues. When lawns have too much water applied, or are watered at the wrong time of day, ants may come inside to get away from what they perceive as a flood.

Outdoor species of cockroaches often rely on moisture dependent fungus on the ground, and moistened leaves which provide for their food supply as they breakdown. Roaches belonging to large populations and sharing a limited food source may move into structures to avoid the competition. When the lawn dries out in the spring or summer they come inside foraging because their food source is disappearing. The key to controlling these and other moisture dependent nuisance pests is to control the moisture. Proper lawn watering is one of the most important factors in control for many insects. It should be limited to what is necessary and appropriate. There is balance with most lawns and diminishing value when using larger amounts of water. After a certain cut-off point the large amounts of water become detrimental by doing more harm than good. From a practical standpoint, large amounts of water do not help with regards to lawn growth and may increase insect activity.

Because of the types of conditions created by moisture and filth, controlling moisture on the property and a consistent sanitation program are necessary in any IPM plan. The responsibility for controlling these conducive conditions should be assigned as early as possible. Homeowners doing treatments on their own take on a large burden, but likewise pest control technicians may assign tasks for homeowners to complete, or to have completed as part of the process. Some issues may be the responsibility of the homeowner, and others the responsibility of the pest control technician. Without clearly defining a recommendation, and assigning a party responsible for the correction, the likelihood of completion is very low, especially when responsibilities should be delegated to several parties.

Mechanical exclusions can be the easiest and most effective control measure for any type of pest. In the case of rats and mice holes on the house are sealed to prevent entry. Many homeowners and even pest technicians use rodenticides near a structure killing rats as they enter which is less effective and the process may go on indefinitely. Excluding rats from the structure would not only limit the use of pesticides, but would also provide a permanent solution, unlike the use of rodenticides which require re-application as they go bad or are eaten. Rodenticides also pose threats to exposed non-target organisms. Most rodenticide active ingredients have a higher mammalian toxicity than similar insecticide active ingredients, and herbicide active ingredients. Rodents are mammals and with increased mammalian toxicity comes increased human toxicity in most cases with a few exceptions in the case of specific modes of action.

Mechanical exclusion is a necessary step in any IPM plan because it lowers the need for chemical application. Exclusion can in many cases can lower pest intrusion with a goal to eliminate pesticide applications all together. Therefore, Exclusion can be used to enhance a chemical application or in many cases can be used to replace it. Types of exclusion can include filling holes, applying caulking, installing weather-stripping, and many other tasks prescribed to lower the levels of pest intrusion through mechanical alterations. Caulking and sealants are applied in many locations, but the most common places include counter tops where water intrusion may attract pests, around pipes where insects or rodents can gain entry and over cracks and crevices where insects may find harborage.

Weather-stripping, door thresholds and door

sweeps can prevent pests from entering a structure through the more obvious areas. These can be easily installed by a pest control technician or a homeowner and can be purchased at a local hardware store. Gaps of 1/64th inches can allow insect entry. Rats can enter through holes as small as 1/2 inch, and mice can enter through holes as small as 1/4 inch. Technicians should always communicate to the customer when and where they find areas of intrusion and should always be ready to do the necessary repairs themselves. As part of an ongoing program, service technicians should always be looking for these areas needing work and communicate these issues to the customer and arrangements should be made to get the necessary repairs and corrections completed.

Non-Pesticidal Repellents

Some pesticides can have repellency to insects or other pests and these repellent pesticides can be used sparingly as part of an appropriate pest management plan. Repellency is at times an amazing feature of those pesticide products, but many non-pesticide compounds have an inherent repellency without a specific chemical mode of action. In many cases cinnamon, nutmeg, and other spices are used as an organic pest control solution. The disadvantage of spices and other compounds with a powerful built-in repellency is the tendency of these products to force insects into unexpected areas, and scatter social insects such as ants, and roaches. Likewise, repellency can cause some insects to lose their appetites, or to avoid areas they may otherwise be expected to go. When these repellent treatments fail, and non-repellent pesticides are applied, they are hindered by the original application of these repellent items. Insects with no appetite will not eat bait products, and insects forced away from treatment areas are not likely to cross the products applied in those areas.

Aversion

One of the worst-case scenarios involving repellency occur when ants are powerfully impacted, and they can be found in multiple areas they would not otherwise be. This problem can be further impacted by larger populations forcefully trapped within a structure because of the repellent substance. Spices, potpourri, and other fragrant household items are not advised as a sole solution for most insects. In the case of ants these common household items should probably be avoided unless specific conditions creating or enhancing the problem are found and corrected first. In many cases ants have been driven inside by the overuse of watering systems along the foundation, or by sprinklers hitting the structure. Using repellency to control ants would be a mistake if these moisture related issues were not corrected first. Ants trapped by the repellency of spices, or other products, and the repellency of the moisture outside could be forced deeper into walls, across new areas inside, or into other unexpected places. This is not better than applying a repellent pesticide to control these ants. The situation would be identical, and control may be gained by correcting the moisture issues alone, and not applying any products.

Castor Oil Treatments for Burrowing Vertebrates

For vertebrate pests such as gophers and moles, castor oil is used as a natural repellent. It has both advantages and disadvantages with animals reacting immediately, but quickly adapting to excessive applications. When it is applied strategically, however it can be very effective. It has been shown to work by making soil undesirable to burrowing animals because it irritates pests enough to push them out of the area. Different store brands of finished products have differing instructions, and techniques may vary. Basic instructions involve a simple straight forward application, but castor oil alone is not likely to penetrate the soil to a high enough degree for proper control to be achieved. When formulated, store brands quite often come with surfactants in the bottle to drive the castor oil deeper into the ground. Natural Yucca Shidegera extract can be purchased online and suits the purpose for do-it-yourself'ers wanting to make their own formulation. One example of a complete gopher repellent product with a surfactant already in the bottle is Chase Mole and Gopher Repellent. The advantage of these castor oil products from an organic standpoint is their basic design. Safety concerns are put to rest by the ultra-low toxicity of diluted castor oil because castor oil at these low rates poses no reasonable threat to any wildlife. Chase brand repellent is even sold and branded as a non-toxic solution by its manufacturer.

Biological control and Bio-Pesticides

Many forms of pest control can be environmentally intrusive, but a major feature of organic pest control is treatments being done in the most environmentally conscious manner. The term organic implies that treatments put forward have some natural methods included. Therefore, Biological control becomes another alternative method in which biological factors are changed to alter the environment. In many cases, living organisms are used to control pests as the primary method for biological control. This includes the use of lady bugs and green lacewings for general pest control, predatory wasps for caterpillars, nematodes for fleas, and even bacteria containing pucks for mosquitos. These treatments can be advantageous in many cases but can also be detrimental when care is not taken with planning services.

Planning your treatments ahead of time is important. While ladybugs have shown great promise with commercial farming, Green lacewings have shown better effectiveness in residential settings. Predatory wasps can be effective but are not selective about prey and can eliminate other beneficial insects. Fleas can now be treated with a lawn spray containing a small species of round worm called nematodes. Beneficial nematodes feed on every stage of the flea life cycle, even those not easily controlled using pesticides. These nematode-containing spray solutions do have a shelf life and should never be applied to areas where pesticides may be present or used in a sprayer not thoroughly cleaned with pesticide residues completely removed.

Chapter 5 Pesticide Applications Within an Organic Pest Management Program

Biorational Pesticides

Traditionally pest control has been accomplished using highly toxic chemicals, but in the last few years it has moved to a more comprehensive approach with the use of integrated pest management (IPM). Many people are unaware of the change, and of the effectiveness of lower toxicity treatments. The average home-owner and the less experienced pest control technician may even believe the higher toxicity products are more likely to be effective. This however is not usually accurate, and in fact the opposite may be true. Highly toxic pesticides can have negative effects on a treatment strategy and low toxicity products can in many cases be a better solution. Therefore, it is the duty of any pest control technician to select the appropriate pesticides for every job, and to base his opinion on multiple factors. Pesticide selection should always have a balance of effectiveness vs safety and necessity vs tolerance with IPM being the primary focus of any organic pest control program.

Pesticides made with naturally occurring residual active ingredients or with specific synthetic compounds believed to be low or non-risk are referred to as biorational pesticides (bio meaning biological and rational meaning sensible). Biorational pesticides include several categories, and they are the general products used in organic pest control. Any one performing pest control services should however be aware of a basic fact; organic pest control and organic pesticides do not equal the same thing. Organic pesticides provide peace of mind to customers who think product selection is important and some of them are appropriate for use as part of an organic pest management service. Organic Pest management, which is more involved than simply using organic pesticides, provides the more viable option with regards to safety and control.

Organic pest management has more to its practice than just the steps of integrated pest management. With organic pest management, pesticide selection is more specific. While IPM alone does not generally require specific classes of pesticides, organic pest management does. Biorational pesticides are the pesticides generally used in organic pest management, and low impact organic pesticides are a category of biorational pesticides. Many pesticides used in organic pest management contain inorganic compounds but are not any less safe than similar organic pesticides. Many inorganic compounds are natural, and some are used with very little alterations from their natural state making them ideal for those seeking a natural solution. Inorganic compounds do not contain carbon, and are in many cases low toxicity, as is the case with many biorational products containing boron. The term organic pesticide also makes no claim as to the toxicity of the product, it's lingering effects, or its safety. Therefore, only low impact organic products are considered biorational.

While many people assume organic means safe, the word organic means something completely different. Organic, when describing pesticides, means the pesticides are made from naturally occurring compounds containing carbon. Inorganic pesticides are the opposite, and while inorganic products do not contain carbon they are also derived from naturally occurring compounds. The pesticides used in organic pest control are always low toxicity, biorational products, and are not always organic products. These include many different products, and formulations with both organic and inorganic compounds. Generally, pesticides containing synthetic active ingredients are not labeled for use in organic pest control sites except for soaps and oils with little or no residual action.

Organic pest control procedures regarding farms and warehouses are defined by the USDA under the NOP (National Organic Program) and pesticides labeled for use in organic farms and warehouses are reviewed by the OMRI (Organic Materials Review Institute). A list of pesticides currently labeled for use in organic production is available online at the OMRI website. When applying pesticides within an organic pest control program, as prescribed by this book, the OMRI list is one source of appropriate products and is the most conservative source regarding biorational chemical standards. Organic pest management within residential structures includes many of the requirements found within warehouses but considers the differing nature of the residential environment. Some pesticides not found in the OMRI list are still appropriate for organic pest management in non-commercial settings, but every product on the OMRI list is appropriate for non-commercial settings when used properly.

Within a proper organic pest management program, the use of pesticides is in many cases limited, and banned in others. For organic food warehouses, residual pesticide use on the inside is always banned, and non-residual pesticides may only be used while the warehouse is closed, and no food products are present. The thought process can be applied to residential properties as well, with limits on treatments, and product labels followed in accordance with the law. While treatments become more time consuming they also become more complete. The IPM standards force pest control technicians to rely heavily on non-chemical methods such as cleaning, sanitizing, and mechanical exclusion. IPM plans always assume pesticide applications as the final step after other non-chemical IPM solutions have been exhausted.

When servicing warehouses and restaurants, there is no specific requirement for a current and accurate sight map. However, it is a good idea when starting and continuing the treatment process to have a sight map created and maintained. Likewise, it is a good idea to create a treatment folder to be left on the property with the labels and safety data sheets of preselected pesticides which may at some point be used on the property. The site map should be kept in this folder with locations of monitoring devices and other important areas. This folder can also be used to keep track of pest problems, and conducive conditions, and what was done to correct them. Service tickets can be placed in the folder, and the folder can in turn be used to keep track of trends, and even to communicate with warehouse staff and management.

Pesticide Safety and LD50

Pesticides can vary in toxicity and while many of them are low in toxicity, the term non-toxic should never be applied to any pesticide compound. The term "safe" is also generally not appropriate because it may imply a level of safety higher than is expected for specific products in non-specific situations. While basic safety can be assumed through a proper application, one could with some effort, possibly find a situation where even the most innocuous products could pose some type of danger, however farfetched the danger might be. In an effort to ensure pesticides with the lowest impact are categorized into their own class, many pesticides have been registered using the term GRAS. GRAS stands for generally regarded as safe. This term implies a level of safety, assumed when all label directions are followed, and all appropriate safety concerns are met. The term GRAS cannot be associated with all pesticides since a pesticide must be specifically registered with the term in mind. It applies to many neem oil based products such as Cirkil and many other plant oil based products.

Pesticide Toxicity is measured in 3 basic ways, with an LD50 for dermal exposure, a different LD50 measurement for ingestion, and an LC50 for inhalation hazard. An LD50 measurement is determined by the amount of pesticide absorbed through the skin, or ingested causing death in 50 percent of a test population. Since the concentration of the amount absorbed is important, a higher number means a lower toxicity. If it takes more to cause death it is less toxic but if takes a smaller amount of it to cause death it is more toxic. As an example, boron (disodium octaborate tetrahydrate) is 2,550 mg/kg in dermal toxicity, while nicotine is 60mg/kg making boron the lower toxicity product. LC50 is on the same scale and the same rules applies where if you have a higher number you have a lower toxicity.

The Case Against SGAR's (Second Generation Anticoagulant Rodenticides)

Second generation anticoagulant rodenticides (SGAR's) are a modern, relatively fast acting type of rodent bait (rat poison). SGAR's have received a lot of bad press in recent years because they have been implicated in the deaths of many non-target animals. The active ingredients found in SGAR's have higher rates of secondary poisoning in most cases. Because of these active ingredients being found in many single use packets, and other over the counter solutions, many homeowners were irresponsible with these products while pest control companies overused them due to the low price at which they were available. They have been widely overused and have been the focus of federal and local government environmental agencies around the world. Those who have made their main control solution IPM based strategies have not been impacted as heavily as those who used the SGAR's regularly because their focus was always on controlling rats through mechanical exclusion, and other types of non-chemical control. When rat poisons are considered a last resort, they are less relied on which leads to better, more environmentally friendly solutions.

To kill rats, trapping is the most reasonable solution. The old fashioned wooden snap traps with the big spring, and metal trigger have worked great for decades and still work great in the new era of traps which are designed to electrocute, or work in other unnecessary ways. Rats are easily caught and killed when properly set traps are placed in their path. The best bait however is not from the commonly used types. Peanut butter is terrible for catching rats because it loses its effectiveness very quickly when it absorbs odors from the air or starts growing mold. Similarly, cheese gets hard very quickly and loses its texture. Rats are very picky with their food selections and will stay away from non-palatable food in many cases. Using a bait which quickly loses its appeal is self-defeating and the likelihood of catching anything is greatly diminished. While cheese and peanut butter are the usual default, they may not be the best choice due to their tendency to absorb odors, and their short freshness period.

All types of anticoagulant baits are generally banned from use in any form within organic food warehouses, and in other similar situations. Those performing residential work for hire, and those controlling rats on their own should also avoid anticoagulant baits within any organic pest control program except for those few Products listed and labeled for use within organic pest control accounts. One method on the exterior to be used as an alternative to poison involves Protecta LP branded bait stations normally designed to hold poison. Instead of poison, non-toxic monitoring baits are used. These non-toxic baits include the brand name Detex, and some others. Once activity levels have surpassed the action threshold, the monitoring bait can be replaced with special T-Rex branded traps designed to be used inside those stations. Likewise, trap boxes, and TinCats (a style of indoor trap with similar brands and types also available) can be used inside of warehouses and other semi-open areas.

Rat poison should be considered as a last option. These poisons have established toxicity levels and modes of action generally harmful to humans and animals. There are alternative baits available of which manufacturers claim to be low risk, and in many cases non-toxic. One product called Ratx by EcoClear Products contains large amounts of sodium which causes rodent death. This product can be used in a relatively safe manner around humans and pets because the toxic dose of sodium is higher in these small animals. The same amount causing death in a rat would not kill a larger animal or person. The company markets their product as non-toxic, but there is always a risk associated with pesticide application, even if the risk is very low.

Plant Based Pesticides, Botanicals, and Other Natural Pesticides.

Many pesticides are plant based, and in even the most synthetic products some plant material is still used to manufacture many products. The most unaltered compounds from natural, living sources are considered organic. Since the term organic is broad, another term used for plant based pesticides, such as organic pyrethrin is "botanical." Botanical pesticides contain plant oils and are often made from flowers. Botanical, however does not always mean organic and organic does not always mean botanical. These terms are in some cases interchangeable, but they don't have the same meaning. A pesticide can be both organic and botanical, organic alone, or botanical alone.

Pyrethrum

The most common pesticide active ingredient made from organic compounds is listed on pesticide labels as pyrethrum. Pyrethrum is made up of 6 compounds including pyrethrin I, pyrethrin II, cinerin I, cinerin II, jasmolin I, and Jasmolin II. Pyrethrum is a sodium channel modulator and blocks nerve signals across the nerve axons by binding to the sodium channel and forcing the sodium channel to stay open. This mode of action is more toxic to insects than humans making it a low risk insecticide. It is derived from the chrysanthemum flower making it a botanical insecticide. Pyrethroids mimic the mode of action found in pyrethrum but are recreations designed in a lab. Natural pyrethrum however, almost always requires a synergist to work successfully unlike the synthetic versions which are functionally designed with efficacy in mind and as their design progressed throughout the years the need for synergists was essentially eliminated. Pyrethrum also breaks down quickly when exposed to light therefore it leaves little to no residual, it can be highly toxic to several non-target species like fish and birds, and it quickly establishes insect resistance with regards to cockroaches, and bed bugs making it ineffective in some cases. Many insects have developed resistance to pyrethrum due to its over use throughout the second half of the 20th century.

Plant Oils

Many plant oils are adapted for pest control use. While natural pyrethrum is a good, botanical solution, many prefer other types of less innocuous plant oils. The more common plant oil products work by binding to octopamine. Octopamine is used in an insect's body as a neurotransmitter to regulate movement, heart rate, and metabolism. Plant oils block octopamine, causing organs and muscles to shut down. Most non-target organisms do not have octopamine receptors in their bodies and therefore active ingredients containing these plant oils are very low toxicity to humans and pose little threat to other vertebrate animals. Many plant oils can also be used near ponds and lakes with almost no damage to fish and other wildlife.

Essentria IC3 is the trade name for one product made from plant oils. The product contains a combination of rosemary, geraniol, and peppermint oil. It comes as a concentrated emulsion and has a surfactant in the bottle. Many other products require the addition of a surfactant or other adjuvant which would increase the efficacy without increasing the toxicity. This product is exempt from EPA registration due to its low toxicity, and the natural components found within. Likewise, the product can be used near open water and is considered safe around fish and other aquatic animals.

Cedar Oil

Another essential plant oil used in insect control is made from cedar. While many consider cedar blocks in their closet to be a natural alternative to moth balls, cedar oil can also be purchased as a ready to use, or concentrated liquid for spray applications. Its uses are similar to the uses for rosemary, and other essential oils, with a few exceptions. One specific use has been with water lilies infested by aphids. The cedar oil was applied to the water lilies for control, and the water body below was not affected. This type of treatment has become a typical use, and cedar oil sprays after a rain can help to keep mosquitos away with very limited danger to the local water table.

Neem Oil

Neem oil is another alternative active ingredient made from organic sources. The natural chemical compounds found in the neem seed provide a killing action through growth regulation and by altering feeding habits. The active ingredient has been shown to work on many garden and lawn insect pests but in recent years it has been adapted for use with bed bugs. Since no insects have shown any resiliency and have not developed any resistance, it is a good alternative to the more commonly used pyrthrins, and pyrethroids. The product Cirkil was developed as a special kind of neem oil for indoor use. Cirkil is cold pressed. Cold pressed chemicals are made without heat which could alter the chemical's composition. This makes it more natural than similar products, such as orange oil, which is heated during its manufacture. Its current usage is solely for bed bugs and has shown promising results in every study. One disadvantage of Cirkil and neem oil in general is the potent odor associated with all neem oil products.

Neem oil when used outside can control multiple plant infesting insects, fungus, and diseases, and it will not affect foragers, or other non-target pests once it is dry. It works primarily on plant feeders such as aphids and thrips. The active ingredient can be used on citrus trees to control citrus thrips and works on plant infesting aphids which are known to infest multiple types of plants. It can be mixed with insecticidal soap products for greater control. The dual mode of action created by mixing the two products will increase efficacy against insects, it will also kill fungus, and will slow the spread of plant viruses.

Orange Oil

The active ingredient d-limonene, which is the active ingredient used in products going by the general name "Orange Oil" is manufactured from citrus peels, and vapor distilled to make the active ingredient. Orange oil's mode of action is believed to be the overstimulation of sensory and motor nerves. Insects lose control of their bodies, and the over stimulation eventually leads to paralysis. Many insects overcome the paralysis instead of dying however. One of the ways this is overcome is by using piperonyl butoxide, which is a synergist designed to enhance the toxic effect of a pesticide. Piperonyl butoxide is the same synergist most often used with pyrethroids and pyrethrins. Orange Oil is generally used for drywood termite control as a local treatment. The product is widely marketed as the alternative to all other forms of termite treatment. However, studies have shown tenting with sulfuryl fluoride is in almost every case a more viable solution (although not an organic one), and borates are the more appropriate organic approach with a higher likelihood of success when locally treating. Orange oil does have an odor, it is a skin and eye irritant, does not have the same killing action as boron based products, and does not leave the same residual as boron (disodium octaborate tetrahydrate). Many companies apply borates in conjunction with orange oil to gain higher levels of control, in fact. The borates in many cases eliminate termite colonies left behind by orange oil.

Surfactants

Surfactants are a type of adjuvant designed to give additional efficacy to a pesticide mixture without adding to the toxicity level, and without enhancing the effect of a specific mode of action which is the case with synergists. Surfactants have several uses, including being used as spreaders. They make water molecules bond together more easily which creates a more even distribution on plant leaves and on ground surfaces. Finished solutions containing surfactants can also penetrate the soil easier making them wetting agents. Wetting agents lower surface tension on soil allowing pesticides to more easily penetrate the soil's surface. Pesticide sprays come in several types of formulations and with certain types such as wettable powders, suspensions, and microencapsulated products soil penetration does not happen as readily as it does for pesticides formulated as emulsions. These products stay at the surface in many cases. Therefore, when applying these pesticides for insects just below the surface of the soil, a surfactant is always recommended. When working with emulsions, the use of surfactants may still be reasonable since it allows for a more even distribution across surfaces.

An extract from the yucca schidigera plant is used as a surfactant in many pesticide formulations. It can be found as a stand-alone product and added to spray mixtures, but it's also an additive in several end use products, and pesticide concentrates with the extract added during manufacturing. The yucca schidigera plant produces its own compounds to manage water more efficiently and the surfactant compounds found within the plant make it perfect for use with pesticides, and fertilizers. When used by itself it can even add to the vitality of plants and correct some soil conditions.

Inorganic Active Ingredient Formulations Used in Organic Pest Control

Boron Based Baits

Because boron is a mineral, and its derivatives are mineral based, boron is considered an inorganic compound. As a result of the dual meanings implied by the term organic, the use of inorganics in organic pest control can be confusing. Boron is naturally occurring and can be used in organic pest control.

Inorganic compounds are the opposite of organic compounds, but those terms have very little connection to organic pest control. A pesticide or active ingredient which is organic in nature is not inherently meant for use within organic pest management, while an inorganic pesticide may be perfectly appropriate. This is because organic pest management is a process, and the general use of organic pesticides does not equate to organic pest control or organic pest management. Natural, low toxicity, inorganic compounds are appropriate for use within an organic pest management plan, while organic pesticides are generally thought of only as those products and active ingredients containing carbon. They may or may not be appropriate based on more specific requirements. Likewise, the narrow definition of organic pesticides does not include any expectation of safety, toxicity, or risk.

Many pesticide spray applications are enhanced using insecticide baits, and many applications can be enhanced very easily with the use of a granular bait in gardens, landscaping, and on turf. Mother Earth Granular Scatter Bait (EPA Registration # 499-515), and other boron based baits are commonly used in conjunction with liquid (spray) applications and can enhance a liquid (spray) application by extending the time period in which control is achieved. Both synthetic and non-synthetic pesticides lose their effectiveness over time and after a period of time have no residual action. The average length of common liquid applications is a few weeks to one month. Mother Earth bait and other boron based granules will last beyond one month and can continue to work in the following months. In seasons, such as summer, control continues when the liquid application's control period ends, and many insects are still active. Mother Earth Granular Scatter bait is not considered an organic pesticide, but its low toxicity, and boron based active ingredients make it perfect for organic pest control and It is OMRI listed for use within an organic pest control program. Boron kills ants by interfering with the conversion of energy inside insect cells, effectively slowing and stopping insect metabolism. Boron has also been shown to kill the dysentery in the gut of some cockroaches which is why boric acid is one option when attempting cockroach control. These modes of action make boron a very low risk active ingredient, and the likelihood of humans or animals being harmed is very low.

Boron Based Liquid Ant Baits

There are several commonly used liquid ant baits made with boron. Some of them contain boric acid, and others contain less complex forms of boron. One of the most common brand names purchased by homeowners is Terro Liquid Ant Bait, while the professional grade counterpart is called Terro PCO liquid ant bait. Terro is a simple formulation containing Sodium Tetraborate Decahydrate (Borax). Terro has proven effective in many cases but has its limitations. When used alone it has the potential when colonies are large, or have begun to show signs of aversion, to lose its effectiveness. Ants will often stop eating as though they have learned the bait is affecting them. Ants however, have varying food sources and with large numbers of workers dying, ants may change food sources to help replenish workers more quickly. Terro is best used to supplement a complete IPM program. It may incorporate the use of other baits, and pesticide sprays, or dusts. Environmental controls are the first step in any IPM program. This includes controlling conducive conditions.

Diatomaceous Earth

Diatomaceous earth is usually applied as a fine powder, and it comes in several forms. It can be a very effective pesticide used in organic pest management. Although it's active ingredient, silica dioxide, is an inorganic compound, it comes from natural sources. The term diatomaceous, refers to fossilized diatoms, which are the algae it is made from. Diatoms have a unique cellular structure with the inclusion of silica which is not common among other types of algae. This silica becomes abrasive when fossilized and absorbs the waxy coating on the outside of an insect's body. As a result, insects usually die of dehydration. It is best applied in small amounts when used within a structure. Too much of it can become repellent and can create new insect issues but when applied properly it can be beneficial. It is easily pushed into cracks and crevices using a dry paint brush, and excess product can be cleaned to prevent insects from over-reacting to its presence.

Chapter 6 Borates for Wood Preservation and Control of Wood Destroying Organisms.

The Basics of Boron

Borates are inorganic chemical compounds containing the mineral boron. These boron-based minerals generally need to be ingested by insects to have a toxic effect. They are almost always stomach poisons and tend to be slow acting. In many insects, borates kill protozoa and dysentery in the gut which interferes with several parts of their metabolism, but mostly stifles their ability to eat and digest food. The most common molecular structure of a boron based active ingredient is disodium-octaborate-tetrahydrate, and this is the standard form. When boron is referred to as a pesticide, this is generally the form being mentioned. This active ingredient is found in products such as Tim-Bor, Board Defense, and Ni-Bor. Another formulation of boron is boric acid.

Boric acid is generally created by adding hydrochloric acid to the form of boron. Either ingredient is appropriate for cockroaches and ants in a dust form, but the standard form of boron mixed with water can be used for wood destroying insects, and organisms. Boron, when applied to wood makes the wood toxic to wood feeding insects. Boron penetrates the dry cellular structure of the wood, and as a result wood destroying insects have their food source compromised. Boron treated wood has shown no resistance to established infestations of drywood termites, therefore they continue to eat despite their food source being poisoned.

The use of boron based products for control of many pests is a low toxicity, long lasting solution. While boron and its derivatives such as boric acid have been widely used for cockroaches, crickets, ants and many other pests, it's application as a wood preservative and protectant has the most efficacy of any of its uses. Because it is mineral based, boron leaves a long-lasting residue on wood, shown to last decades, and in some cases, up to a hundred years. Boron is considered low toxicity and many boron products are registered as GRAS (generally regarded as safe).

Boron in Nature

The element Boron, which is part of every boron based pesticide is a micronutrient found in plants. It is required for pollination, germination, and plant growth. Without it plants could not survive. Boron can be found in all parts of plants including stems, leaves, and fruit and is found in plant derived foods, and in many cases, it is also found in the water we drink. The human body is well adapted to boron consumption making it one of the least toxic pesticides available. Boron is used by nature in differing amounts within plants and therefore the line between too little boron, and too much near plants is often very fine. Therefore, applying boron to plants should be avoided, and care should be taken with drift. Plants over-exposed to boron will often show signs of phytotoxicity, which is made obvious by what appears to be burnt and damaged foliage and leaves.

Boron is used as a wood preservative and protectant in the mineral form of disodium octaborate tetrahydrate. It generally comes from the manufacturer dry, but in some cases, it's formulated differently for specific uses such as with products containing glycol which allows for the faster absorption of the product into wood. In its concentrated from its generally not applied, and should usually be mixed with water prior to application, except for a few products and situations. For dry forms at a 98 percent active ingredient, the technical grade is mixed 1lb per gallon of water to achieve a 10 percent mixture in the final solution. With liquid concentrates such as BoraCare the products are generally mixed to different levels with a 1 to 1 ratio of product with water forming a 23 percent solution, and 2 to 1 mixture forming a 16 percent solution.

Mixing borate products is best done with powered mixing equipment such as electric drills and attached paints mixers, or other mixing equipment. While BoraCare and the other diffusible borates mix more easily with water, the dried products such as Tim-Bor and Board Defense are very hard to mix with water and will settle when not mixed properly. Properly mixed boron should be clear, to yellowish color, with a slight glisten. Improperly mixed boron will appear cloudy or white in color.

Once liquid boron is thoroughly mixed it can be applied to unpainted wood members before of after the wood members become part of the structure. The most effective method for many houses is a spray application applied while houses are still in the critical framing stage. At this point, most, if not all the wood within a structure is exposed. These pre-treatments vaccinate a house for termites and can limit the necessity of further treatments for several years.

As termites feed on the wood treated they consume boron embedded in the wood. The boron will then kill the protozoa in their digestive tract which is necessary for proper digestion of wood, and without the protozoa, termites die. Studies done have shown boron diffusion increasing exponentially over time. The rate at which it is absorbed is determined by the moisture content of the wood, relative temperatures, humidity and other factors which might limit or enhance its absorption rate.

Boron as a repellent

Boron can also be used as a repellent for subterranean and formosan termites. These types of termites are known to avoid boron treated areas and will not usually cross boron treated wood members to get to untreated ones. This application includes treatment of the wood members but will often also include the concrete foundations and other surfaces termites may cross to reach the structure. When termites are pre-existing in wood and the wood is then treated however, studies have shown they will continue to feed into the treated areas. This is especially helpful in the case of drywood termites, which live solely in the wood they feed on and continue to eat even though their food source has been compromised. They ignore the boron and slowly work their way through the treated area, eventually causing colony collapse.

Product Specifications

Tim-Bor and other dry boron products and generics are mixed at a rate of 1 pound of product to 1 gallon of water. This ensures a mixture ratio of 10% active ingredient. The 10% active ingredient percentage is the appropriate level for most applications. Because boron forms a dilution, applicators should use warm water when mixing dry boron products. This will allow an easier mixture to occur. Likewise, mixed boron should not be stored overnight since it can settle, and it may be hard to re-agitate. This complicates the treatment by making the active ingredient percentage an unknown amount and making the post-treatment cleanup a daunting task. If mixed boron is ever stored, it should be mixed again prior to application. Boron solutions are usually applied at rate of 1 gallon per 200 square ft of surface area. When using the dry boron mix two applications are recommended to all exposed, unpainted wood members with complete coverage in mind.

Another type of borate which has some more advanced uses is the glycol containing borate concentrates such as Bora-Care. Bora- care comes in a liquid concentrate and is mixed with water, forming a similar dilution to what is found with the dry concentrates. Bora-Care however can be more easily mixed at higher concentrations, can be used for additional pests, and in some cases, it is labeled for mold. Bora-Care and other similar glycol containing borates are usually mixed with a 1:1 ratio of technical grade product, and water giving us a final mixed active ingredient percentage of 23%. In the case of RTU products such as Jecta the product is used at the technical grade (no water added). Jecta comes in a caulking tube and is applied using a caulking gun. It is applied to individual wood members by a drill and inject method. The high concentration and fast absorption make Jecta perfect for rafter tails, patio posts, and other outdoor structural pieces. The active ingredient percentage for Jecta is 40%, which is identical to Bora-Care at its technical grade. The glycol in the mixture increases the absorption rate of boron. These products have their advantages especially when treating in areas of mold, or wood destroying fungi. While dry boron formulated with water at 10% is considered appropriate in most cases, these glycol-containing products are considered more effective. The disadvantage however is the price, and in the case of Jecta the application method is time consuming. Likewise, glycol containing borates may not be considered organic, and although they may be low in toxicity, they may not fall under any Standard for GRAS registration.

While boron works great as topical treatment during the framing stage it must be applied differently to finished houses. Boron foam applications to windows, and doors can aid in termite prevention, in addition to the topical treatment of attics and crawlspaces. Boron foam can also be injected into wall voids, and other important areas. Holes are usually made in the drywall to allow the insertion of an injector tube from a foam machine and in many cases companies looking to treat the lower wall voids will drill holes along the lower level wall studs to apply boron as a preventative treatment. As a rule, the wall studs within a structure are approximately 16 inches apart. If mixed properly the boron foam should treat large portions of wall void, and other framing by expanding into treatment areas. Boron foam also works well as a local treatment method for individual pieces of wood or several pieces in question. One major advantage of boron over other local treatment products is its extended functionality through advanced uses. It can be injected into or applied topically onto surrounding wood members for more complete control. Most other products can only be injected, but because of boron's tendency to move throughout wood, wood can be treated from inside the wood member by injection, or the surface with a topical application. In many cases both options are available.

Wall Void Treatment Examples

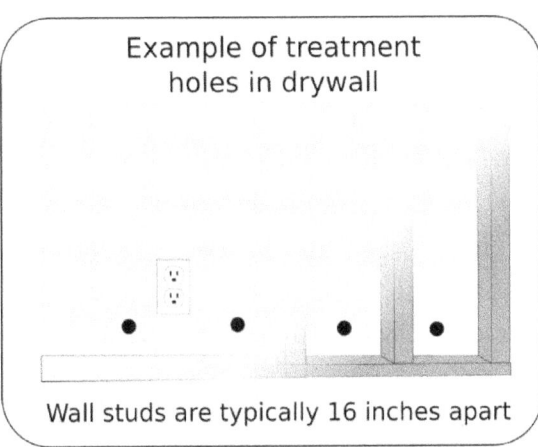

Example of treatment holes in drywall

Wall studs are typically 16 inches apart

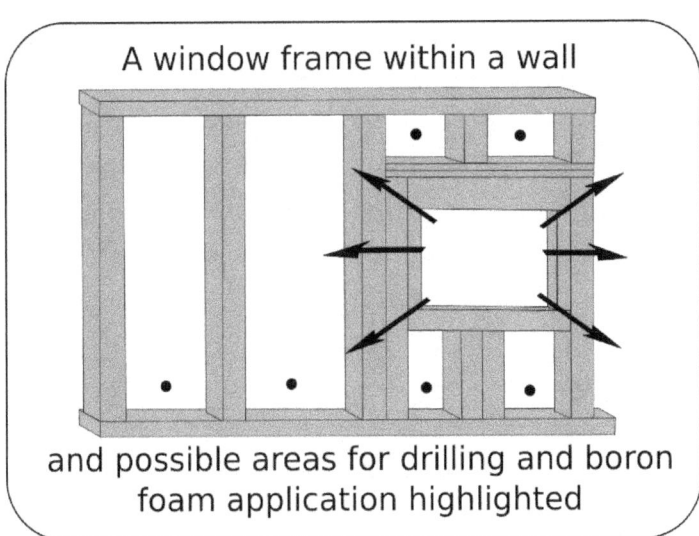

A window frame within a wall

and possible areas for drilling and boron foam application highlighted

Other Considerations with Boron Wood Applications

Because boron is water soluble and moisture affects its absorption, boron treated wood should be kept away from constant sources of water and treated wood in an outside location should have a water-resistant finish. As an example, outdoor decking should be treated with proper a water seal such as Thompson's after the boron has been applied and given time to dry. Treated, unprotected wood should not be in contact with the ground or left in extreme weather conditions for an extended period because over time the boron will be lost.

One effect of boron on wood, not commonly discussed or explained well is the flame retardant nature of boron treated wood. As boron is absorbed into wood a new compound is produced called boron oxide. This boron oxide makes boron treated wood incapable of supporting a flame. 2 applications of boron at 10 percent or more active ingredient are sufficient for houses in the framing stage to limit house fires and the spreading of fire throughout a structure.

Because boron crystalizes, sometimes even inside of application equipment and hoses, it can cause multiple issues and negatively affect applications. Therefore, cleaning of all application equipment should be done regularly and after every application.

Spray tanks should be emptied, and all hoses cleared of boron mixture leaving only clean water in the hoses and the spray tank. Boron should never be mixed with any other end use pesticide without a specific reason for it, such as a label recommendation.

While some treatments may include the application of boron as a topical wood treatment and a second product as a soil treatment, like in the case of subterranean termites, technicians need to be careful of boron residue in the spray tank while applying termiticides to the soil. Boron residues within a finished termiticide formulation intended for the soil can negatively affect treatments and are strictly forbidden on several common termiticide labels.

Chapters 7 Strategies and Case Studies

Pantry Moths

A few years back I was called out to an apartment to investigate flying insects seen throughout an apartment unit. As the inspector I was given an opportunity to service this customer with very special requests. She wanted the least toxic method for controlling the situation, which was later determined to be Indian meal moths. She requested the lowest toxicity pesticides to be used because even rosemary oil and other liquid applications were simply "too much." I therefore focused my treatment primarily on the principals of Integrated Pest Management. The approach started with a thorough search through the cabinets in her kitchen, paying particular attention to the common foods which might attract these external feeders. After all the food in the cupboard was removed and inspected, the infested food items were discarded.

While the cupboard was empty a shop vacuum was used to clean each shelf. Using a crevice tool on the vacuum I paid specific attention to the cracks and crevices where shelves come together, hinges attach, and any other significant areas where moths might have laid eggs. Vacuuming and removing all live specimens was important and therefore done meticulously. Finally, the cracks and crevices were treated using a bellows style hand duster (a tool used in pest control to apply dry products) and NIC 325, a low toxicity biorational pesticide dust. it was forced into the cracks and crevices using a bellows duster and not applied as a broadcast application.

With flying insects, where you see them is generally a destination and not necessarily a source. This was one of the most important factors I had to keep in mind when working. Removing infested food then vacuuming and treating cracks and crevices is also very important in these instances. If the sources aren't corrected, then they will continue to resupply the destinations. Indian Meal moths, and most pantry moths are attracted to light. Adults can usually be seen throughout a structure, and quite often, far from their source because of this tendency to seek out light sources.

In this case, one treatment was sufficient, and the house was monitored using pheromone traps over the course of the next 6 months. No new moth infestations were discovered over the monitoring period, and the customer changed her shopping habits to exclude some items possibly infested at the time of purchase, and she began inspecting packages at the store before purchasing.

Other biorational products which might be used similarly are Eco Smart branded products containing rosemary oil and are sold over the counter. For professionals, there are more advanced products such as those containing hexa-hydroxyl. For pest control professionals, many products come in concentrated forms and the wettable powder products such as ECO PCO-WPX can leave an obvious film behind and their use should be avoided in conspicuous areas.

Ants

Ants have always been a harder pest to control, especially when including organic pesticides in the control plan. The disadvantage of organic liquid sprays is the repellency levels of these usually plant and oil based products. When a pesticide is highly repellent it creates many unpredictable issues, and pesticides of low repellency may be just as bad with regards to several species of ants. This forces organic minded individuals to focus on other aspects of the protocol and to have a higher level of selectiveness with regards to pesticides.

Dripping faucets, inside and out can add to the ant problem, while food sources around the interior and exterior may exist. Some species of ants prefer sweets. Therefore, you can find them inside around spilled drinks, or candy left on a table. Careful thought should be put into cleaning and making the interior less desirable. On the outside of the house, sweet desiring ants may be found on plants covered with honeydew from primary feeders such as aphids.

In many cases ants entering a structure can be problematic. An ant problem within a structure can differ in severity from miniscule, to completely overwhelming. The most basic organic solutions include physical alterations, cleaning, and controlling moisture. Control measures continue with more severe environmental alterations such as applying caulking and trimming or removing plants in contact with the structure.

In multiple cases ant problems have been connected to moisture issues. In one case, a pest control customer with argentine ants in every room of his house corrected the issue by doing no more than changing his lawn watering cycle. In fact, this customer in the desert area of southern California had been watering his lawn 3 times every day. His idea of extra water was an attempt to compensate for the desert heat. As I approached the property for inspection I noted obvious signs of lawn fungus resulting from the excessive moisture and immediately determined the cause of his ant issue. 3 times per week seemed more appropriate, and around 5 in the morning was the adequate time. After the corrections were made, the ant problem eventually came under control.

When pesticides do become appropriate, organic oils, and plant based sprays are the least effective. One of my pest control customers in Ventura County, California had called me to evaluate her ant situation. She had in the previous week paid another company to "spray organic pesticides" around her house. The next day she found ants inside the structure despite having no sign of ants the night before. After reviewing the products used I determined the foundation of the structure had been treated using a product containing rosemary oil.

Several more issues stood out to me about the previous treatment. I first made note of the interior treatment, which had gone undone. An interior treatment should always be done when a repellent pesticide such as rosemary oil is used on the exterior. Repellent pesticides have a particular effect on ants and can force them into a structure. Ants are easily trapped by these products, and if they are between the repellent barrier and the structure of which it is applied to protect, the ants have no choice but to move inside.

I provided her with detailed information regarding why her service had failed, with the misapplication of organic pesticides being the primary reason and explained what should have been done. A proper treatment may have involved the use of plant oils inside, where no ants where present, and the application of granular insect baits labeled for organic pest control on the outside. The solution in her case was time, I estimated somewhere around two weeks from the date of the original treatment the ants would start to leave the structure as the barrier lost its potency. By the two-week mark, all of the ants were gone with no further chemical application.

Granular ant baits, when available may be an option. There are several types with high efficacy and if pest tolerance levels are set reasonably, they can accomplish an adequate level of control. The active ingredients range from minerals, algae, bacteria, bio-pesticides, and traditional organic ingredients like pyrethrin, and d-limonene. These baits outside may be used alone, and some of them are labeled for use in void areas such as crawlspaces and attics. Their use should always be combined with an interior treatment if the ant species noted travels both inside and out such as argentine ants, and the ants are shown to be a nuisance inside. Interior baiting is generally done with liquid baits such as terro, but other types of bait labeled for organic use can be found, and when possible interior and exterior areas should be inspected for issues in need of correction.

Unknown Bites

Pest control companies receive a tremendous amount of calls each month with regards to insect bites. Many of those calls frequently turn out to be something else entirely. On the other hand, bed bugs are a real problem and fleas can quite often become a recurring pest. While the first call to action for many people, especially those feeling overly irritated, may be pesticide applications, it is not recommended. Applications done before identifying a problem is an ill-advised solution. It defeats the purpose of organic pest control by ignoring the most important step in controlling the problem which is properly identifying what it is. A misidentified problem might never get solved because the wrong solution might always get used. In some cases, a problem might control itself naturally, causing a false sense of faith in the process. This is regardless of the improper treatment methods, and when these pests return, further treatments may be ineffective with no understanding of how to correct the problem. This can lead to a forced control solution which may involve higher doses of the wrong product, and larger amounts of chemical exposure. Properly identifying the problem is the most important step in any pest management plan.

On several occasions I have come to houses as a professional inspector to look for biting insects and discovered an over application of pesticides which may have been the cause of the problem, an exasperation of the problem, or had created a new problem once the original problem had been corrected. On many occasions, I found the over application of dust based products specifically exasperating issues.

Carbaryl dust, boric acid, and silica has been directly related to skin irritations, and breathing issues. On one occasion when I was new to the industry I came to house to inspect and the entire floor of an apartment unit was covered from one end of the unit to the other with boric acid. The customer informed me of a biting insect problem she had been dealing with for a while, and even though the boric acid had temporarily corrected the problem the situation had since become worse than ever. Even with my limited experience I could see several issues with her situation. No insect could reasonably trek across her floor and make it to her without crossing the overwhelming amounts of the chemical compound. If she had started with an insect problem, the possibility of her own treatment brining the pests under control seems plausible, however boric acid is a skin irritant as noted on its safety data sheet and could have been the sole source of her issues after the application. The proper treatment for her unit involved removing the boric acid and installing pest monitors. Her condition cleared up without any other treatments, and no biting insects were ever found in the monitors.

Delusional parasitosis is a condition found in a large amount of people defined by the irrational belief of a parasite on, or in their bodies. From my time working in the pest control industry I've had a few confirmed cases of this condition, and several non-confirmed but reasonable cases. It is important for pest professionals to be delicate when talking with clients who might have this condition, and to not offer a diagnosis in any case. Professional pest control treatments should never commence unless a pest has been identified and a proper treatment prescribed. Those non-professionals performing treatments such as homeowners, and property managers should take the same precaution for the sake of avoiding new issues. If, after a thorough inspection, no insect or pest is found, those affected, should seek medical, and possibly psychological care. Professional pest technicians should ask clients with no confirmed issue to get a second opinion from a dermatologist or other medical professional with the ability to properly identify skin problems. If the medical professionals believe the issue to be delusional parasitosis then they can recommend psychiatric care. Pest control technicians have no standing to make such recommendations, but having a client confirm their own skin condition and have a diagnosis from a doctor moves the burden of proof to the client. With treatments not done until a pest is identified, further unnecessary treatments are avoided which might go on indefinitely because those affected would continue to believe they have issues when they do not, or they may have other issues entirely.

Very often those affected will perform their own research on the Internet and find others with similar conditions. The modern internet however, seems to spread as much bad information, as good. I've had several cases over the years of customers claiming to have "no-see-ems." "No-see-ems" as explained by many people in chat rooms, and message boards, are small invisible insects, and according to most of the "experts" on the internet, pest control companies have terrible issues with controlling them. At the time of writing of this book, no entomologist, dermatologist, or vector control expert has confirmed the existence of "no-see-ems."

Dermestid Beetles

In many instances contact dermatitis is found to be the cause of skin irritations. Because contact dermatitis can be confused with bug bites, pest professionals will quite often be called to assess the situation. Contact dermatitis has many contributing factors which may include soaps, deodorants, dust, dust mites, and other allergens such as shed hairs from beetles in the Dermestid family. The Dermestid family includes carpet and hide beetles along with many other common types. The hairs of the carpet beetle larvae break off and cause skin irritations. This is a natural defense mechanism, but the chemical within the hairs causing theses skin irritations remains in those hairs long after Dermestid beetles become adults and the larval casings with these hairs are shed. When carpet beetle larvae are disturbed they will often release these small hairs into the environment, but likewise the casings left behind from matured carpet beetles will break apart and release these hairs. Therefore, control of the pest problem may include a large amount of cleaning to remove the dead insects and their shed casings. In fact, many treatments are done with no chemical application, and only cleaning.

Every carpet beetle treatment should start with vacuuming before anything else. This includes the darkest areas of the house, inside of closets, under and behind things, and the more open areas like the carpet and the furniture. Infested clothes, and garments should be thrown away or cleaned using hot water and soap. Organic pesticides labeled for carpet beetles can be used, but only when necessary. Areas can be sanitized to neutralize the allergens, but allergens can never be 100 removed from a structure once they are there.

Carpet beetles entering a house through doors and windows from the outside is very common. Flowering plants close to the structure are almost always the source. On multiple occasions, I've found them on wisteria, bougainvillea, and morning glories. Flowering plants on a house, especially flowering vines, have the highest likelihood of sending carpet beetles into a house.

On one specific occasion, I was called out because a customer's suits were being eaten by carpet beetles.

After searching the entire house, I found them only in the specific closet where the suits were kept. During my inspection, I noted a bougainvillea on the left side of the house adjacent to the closet. No carpet beetles were found in the large plant itself, but upon further inspection I noticed fallen flower pedals pushed into the substructure. I went underneath and discovered several years' worth of flower pedals the hired gardeners had been pushing into the area with their leaf blowers. This large pile of pedals was directly under the closet and the source of their problem. Removal of the debris was the solution to their problem and if the tree and its fallen pedals had been cared for, the problem would have never existed.

Dust Mites

Dust mites can cause similar symptoms and can also be confused with bed bugs. Their droppings are small and while they cannot be seen by the human eye, they are seen by the human immune system which overreacts to their presence. The most important factor in dust mite control is vacuuming and cleaning. Dust mites accumulate where dust is found. Likewise, they are in many areas you find no evidence. Cleaning all areas thoroughly is necessary for control and bedrooms sheets, and bedding should be washed regularly. Amazing-Solutions.com has a product called Easy Air which comes as a general use spray, or as a laundry rinse. It kills dust mites, and de-activates allergens for several weeks. Professional pest control companies may space spray a room to treat bedding and furniture with a product called Steri-Fab.

Springtails

In many cases, springtails can be a major nuisance. They can be found in some cases migrating into houses in large numbers and finding their source can be difficult. Identifying a reason for their presence can also be difficult, but controlling springtails almost always involves correcting moisture issues on a property. It can be as simple as stopped drainage, sprinklers pointed at the foundation of a structure, or something more complicated like rain gutters packed with fallen leaves not cleared for several seasons.

In one specific case I had a customer with springtails in her upstairs bedroom. She had applied over the counter pesticides to the interior areas of the room and had no success in controlling these persistent pests. When I arrived on the property I started by focusing my inspection first on the common reasons why most people have them. While

I noted the outside of the property was very wet and shaded, no springtails were ever noted on the lower story of the house. I had made a note at the beginning of my inspection of the second story bedroom being an addition, and while this may not seem important at first, it made the difference in solving her problem.

As I looked across the roofline, the roof of her upstairs bedroom had no pitch, which contrasted with the other areas of the house. I took the next reasonable step which was to set up a ladder and get on the roof. I immediately noticed the section of roof over her bedroom was flat with no pitch just as I had suspected, and it had a large lip around the edge making it impossible to see the area from ground level. When I climbed onto the roof I sunk knee deep into wet leaves and pine needles which accumulated over several years, and as I sifted through them I found live springtails confirming my suspicions. The problem was created on the roof and was then spreading into the house.

The treatment involved a thorough cleaning of the roof, and the removal of all leaf debris. This was the only form of treatment needed and no pesticide was ever required on the property to control the springtails. The treatment exemplified organic pest control, and no one could argue this type of treatment as not being green. The ongoing maintenance on the property is keeping the roof as clean and dry as possible. If the roof is kept clear, then her issue will most likely never return.

While in many cases cleaning, removing leaf debris, and other environmental changes are appropriate, pesticides may still be required to correct many tedious issues. In some cases, moisture issues can be a constant problem due to the local weather patterns, too much shade, and slow adaptation to seasonal changes from those on a property. Oil based sprays can sometimes be the most appropriate because they attach more easily to smaller insects like springtails and are more quickly absorbed through insect cuticles as compared to wettable powders. This makes plant and essential oils perfect for these insects. The disadvantage of using biorational pesticides in the conditions necessary for springtail infestations is that many of these products break down quickly and need to be reapplied more often because of high moisture levels in the soil and on the ground.

Fleas

On several occasions I've been asked to customer's houses for flea issues and either chemical applications were unwanted or had already failed. In many of these cases a host animal was found to be carrying and dropping most of the fleas. When dealing with a family pet the course of action is generally easy. Family pets can be dipped, collared, or treated and bedding treated, cleaned or thrown away. In extreme cases, insecticide treatments can be necessary. In a few cases over the years however, the problem came from wild and feral animals. These cases tended to be very tricky because wild animals are sly and most affected homeowners did not even realize they were present.

The solution in those cases was first the removal of the wild animals. Surprises were always expected like in one case where several families of raccoons were living on the side of a house. On another occasion it was just one raccoon, but it had found harborage under a converted deck which was built on top of and had now become an extension of the structure with finished walls, and a ceiling in an effort to make the kitchen larger. Control in both cases was complicated, but after about 2 weeks from when the animals were removed, the problem subsided. The one non-chemical control measure always is necessary is the limiting of access allowed to wild animals on the property. In many cases I find these animals making homes in accumulations of human items, storage sheds, and very often substructures of homes. Keeping yards clear of storage, sheds locked, and substructure access points properly sealed is a necessity in preventing and controlling infestations.

Adult fleas can be susceptible to organic pest control sprays containing pyrethrins, but flea larvae and eggs are generally not susceptible. Complete flea control is achieved by performing a process and not just applying liquid pesticides. Vacuuming every day and forcing fleas out of their pupal cocoon is an important step and never be excluded. Full control can be achieved in many cases within about 10 days. This time frame is dependent on several factors however. 1. Have nesting areas been treated, cleaned, removed or otherwise controlled? 2. have possible re-infestation sources been adequately inspected, prepared, treated when necessary? Dogs, cats, birds, other pets, wild animals, rodent pests, animal nesting, and pet bedding all should be considered. 3. have any retreatments been done within the ten days that may help or hinder the process? Some organic pesticides, particularly pyrethrin, can be formulated to have a short residual of Around 24 hours, and others are formulated to last longer. Fast acting pesticides need to be reapplied more often than slower acting residual pesticides.

German Cockroaches

For a few years I worked for a larger, family owned, nationally available pest control company. They had a focus on lower toxicity pest treatments and they worked hard at developing non-invasive complete strategies. As a primary control method for German cockroaches they primarily focused on the use of a boron dust product called Nibor-d. They had studies to support their claims of its effectiveness and their process had been perfected over several decades.

German cockroaches can be one of the hardest pests to get rid of and many German cockroach treatments fail. With this in mind, I worked through their process identifying what steps made it different from other treatment types. The one most obvious step was to clean, and vacuum all of the infested and possibly infested areas. Eggs, shed skins, and feces removed through vacuuming, and sanitizing could have sustained the roach population indefinitely, but likewise, live cockroaches in trash bags and vacuums taken from the property were removed as well. If more debris, and cockroaches were removed in the early treatment stages, it was more likely to be a successful treatment. Roach feces and sputum will very often need to be removed with a solvent like simple green, or some other type of cleaner. These byproducts of the cockroach infestation give sustenance to cockroach populations, and treatments are affected by their presence.

After vacuuming and cleaning, the treatment continued with the application of boron dust as the only chemical compound used. Nibor-d and other similar dusts are natural compounds, but when cockroaches eat them they become fatally compromised. Applying the dust right where insects hide is the most effective way to force contact with the product, which then increases the likelihood of the insects receiving a lethal dose. It should be used as crack and crevice treatment and as a void treatment. Cabinet hinges should be treated, gaps between pantry shelves, and other tight areas where cockroaches may hide.

Chapter 8 Conclusion

From the beginning, I hope to have shown that "green" does not mean the same thing as "organic." I have also shown the techniques, strategies, and even the terminologies to prove organic pest management a more appropriate title for the process, as opposed to organic pest control. Because Integrated Pest Management has an imposed limit with regards to pesticides, and Integrated pest management is the cornerstone of organic pest management, those performing treatments limit exposures of pesticides for themselves, those around them, and every living thing within the local environment. The limiting of pesticides also makes apparent that in some cases treatments are done with no pesticides at all. When pesticide use is limited, the associated risks are lowered, but when pesticides are not used in any manner, all concern regarding pesticide toxicity is removed, including the inherent anxiety associated with even the most innocuous pesticides.

The largest concern in the last decade has come from the real concerns associated with the overuse of specific pesticides and chemical families. These concerns can be associated with real damage done by these pesticides, and the psychological, or political concerns associated with the use of increasing amounts of these products. Specific examples include the use of pyrethroids in the state of California, which is heavily restricted due to water table concerns, and fipronil in New York for the same reasons. The largest concern in Washington state concerning pesticides has been over the use of neonicotinoids and damage to bee colonies. These products come from the most commonly used chemical families. The level of attention they are given is not a coincidence and if their uses were more limited, it is likely they would be far less scrutinized. This should be a reminder of the most basic point made throughout this book. Pesticide use should be limited, and it should be the last resort solution to a problem most likely unresolved otherwise. Likewise, any pesticide can be overused, and switching to organic pesticides is not a solution by itself. Organic pest management is a process, and when followed properly it will limit the environmental, social, and psychological damages associated with pesticide application.

Organic pest management as a standard is reasonable and effective. Because pesticide use is limited, and in many cases non-chemical methods are a large part of the treatment, costs can be lowered, except in a few cases. For some pests, weather stripping, screens, and other forms of mechanical exclusion may have a large upfront cost. Exclusion work however, does last indefinitely while pesticides need to be re-applied. Exclusion work may pay for itself in a very short period. Screens, caulking, sealants, and other repairs have a long-life span, while pesticides generally have a short life span.

Organic pest control does have its limitations. The use of organic pesticides as one example may have little effect on certain pest problems, and the overall process may take longer without synthetic pesticides. Immediate relief is sometimes wanted and warranted. With pests like bed bugs the benefits can outweigh the risks when using synthetic products. Sometimes fumigations of entire structures are warranted when the costs, and relative risks are considered, and no better options are available. This creates the need for maintenance, and ongoing diligence regarding pest control. A property effectively monitored, where IPM is thought of as ongoing process will always be less likely to reach a point where more toxic options are needed, and more things need to be considered.

There are many considerations when working with pesticides. While organic pest management may limit the concern, there are still many factors potentially affecting treatments. The most common concerns when applying pesticides regardless of classification are drift, and run-off. Drift and run-off occur when pesticides are carried into non-treatment areas by the wind or moving water. These non-treatment areas may include adjacent properties to the one being treated, waterways, and other areas treatment might be considered undesirable. In recent years lawyers have essentially created a new form of litigation regarding drift, and runoff. Many lawyers refer to this movement of pesticides as illegal chemical trespass, and illegal chemical trespass occurs when the site at which pesticides have moved is found to be undesirable, or unwanted, and the applicator has broken the law. Pesticides throughout the United States have been limited by law to the those uses listed on their labels, and without a government exemption other uses and misuses are violations of law. Because of "Global harmonization," the laws, regulations, and labels of pesticides are becoming synchronized throughout the civilized world and illegal chemical trespass is becoming more prosecuted than it ever has in the past.

Another large consideration when working in pest control is the odor of the products used. This includes pesticides, but also glues, caulking and other gas releasing products or products with an odor. Modern synthetic pesticide compounds such as neonicotinoids, and the newest pyrethroids generally have very low odors. These synthetic products however, can change an organic plan into a non-organic plan with one use. Organic pesticides on the other hand, like pyrethrum and neem oil do have an odor. Odor becomes a concern in many cases and should be considered when determining a treatment process. This is one of the many reasons why IPM is so prominent in organic pest control. Many organic pesticide sprays have odors and limiting pesticide use also means limiting the unwanted factors associated with pesticides such as smells. Baits and dusts such as boron based granules, and diatomaceous earth are a great alternative to liquid sprays which might have an odor, but even when using them, control should begin with environmental alterations.

Our society has put a large emphasis on "going green" and being an environmentally conscious citizen is now an important part of the social contract. Organic pest management is one of the many ways industries have adapted to this new stream of thought pushing the world forward. Organic pest management is an alternative to standard methods, but there is no reason organic pest management can't be effective in virtually every case. Organic options combine with Integrated Pest Management in such a way that in many cases, organic pest management produces the greatest result, and it has within its process the steps for limiting environmental impact.

Glossary

Action Threshold: The point at which pests have surpassed the tolerance level and treatments may be warranted. *(See also: tolerance level)*

Aversion: *(see also repellency)* An insect's natural resistance to an area in which chemicals have been applied, or some other deterrent may be present. Ants might stay away from areas with too much moisture, or a lack thereof. Chemical and non-chemical repellents can drive insects away from specific areas. Rats can also suffer from bait aversion if rodenticides are over used or misapplied, and trap aversion as a result of misplaced, and misused traps.

Biopesticides: Pesticides made with biological control based modes of action. They include bacteria based products for controlling insects, beneficial nematodes and other similar pesticides.

Biorational Pesticides: The low toxicity pesticides used in organic pest control. These include 1.) Organic/botanical pesticides such as plant oils 2.) Inorganic pesticides such as boron and diatomaceous earth 3.) Synthetic compounds with no little or no residual action such as soaps and oils and 4.) Bio-pesticides such as nematode sprays and baits containing bacteria.

Boron: An element on the periodic table of elements which is the base of many pesticide compounds. When the term boron is used in pest control, generally the base form of disodium-octaborate-tetrahydrate is implied, and it is the ingredient used to treat and prevent termites, as well as ants, and cockroaches. Boric acid is a more complex compound formed using basic boron, and other compounds created through a chemical reaction. Boron based pesticides are manufactured as dusts, baits, and liquid solutions.

Botanical: A term used to describe plant based products. Many pesticides are botanical such as plant oils, and pyrethrins. An active ingredient may be botanical, but the end product and final formulation may or not be.

Broadcast Treatments: Less selective pesticide applications covering broad areas; as opposed to spot treatments.

Crack and Crevice Treatments: These treatments are more precise than broadcast applications and can be more effective because pesticides are placed directly into insect hiding areas such as cracks and crevices missed by broadcast treatments. Crack and crevice applications also limit human exposure and typically lower the amount of pesticide used when compared to broadcast treatments.

Flushing Agents: *(see also repellency)* These immediate action pesticides are designed to force insects out of hiding when applied. They can however, force insects deeper into hiding if used improperly.

Formulation: The type of pesticide as defined by its physical form. Dry pesticides may come in a granular or dust formulation, but liquids might be wettable powders, flowables, emulsions, solubles, or other types. A formulation is different than a mode of action.

Global harmonization: The worldwide standard for pesticide labels and safety data sheets. The current system relies on pictographs to make note of hazards since labels may not always be in the language of everyone possibly affected by a product. Labels are still considered the law and pesticide applicators should not use a product unless they completely understand its label.

GRAS (Generally regarded as safe): This term applies to ultra-low impact products considered safe when used appropriately. These products have little risk of damage to people, pets, and the environment if all label restrictions are followed.

Insecticide: A specialized class of pesticides used to control or mitigate insects.

IPM (integrated Pest Management): A pest Management Approach containing steps and procedures with basic standards such as limited pesticide use, and limited environmental impact whenever possible. It is an important part of an organic pest control (management) program because it provides the standards.

IPM Plan: a detailed plan of action setting the basis for control on a specific property. This plan is prepared before any control work is done, and establishes strategies, and boundaries to be followed.

LC50: The established toxicity level of pesticides determined by inhalation. A lower LC50 means more toxicity for a pesticide.

LD50: The established toxicity of a pesticide determined by absorption. Many pesticides will have a dermal LD50 for skin absorption rates and an ocular LD50 for the lethal dose absorbed through the eyes. A lower LD50 means more toxicity for a product.

Mode of Action: Every pesticide compound works in a specific way. The manner in which it affects pests is referred to as the mode of action. Mode of actions can be specific by defining how they impact pests on the cellular level such as axonic poisons, and descriptions can be more vague such as stomach poisons and desiccants.

NOP: The National Organic Program is an agency under the USDA working to determine the most stringent standards for organic pest management within organic food warehouses, farms, and other locations related to organic food production.

OMRI – Organic Materials Review Institute: A group working to review and list chemicals, and products used within organic food warehouses, farms, and other locations at the time of organic food and organic commodity production. This includes fertilizers, pesticides, livestock health care products, processing aids, and a number of other products organic farmers and processors rely on every day.

Octopamine: A chemical neurotransmitter found within insects affected by plant oils and some types of synthetic pesticides.

Organic Pesticides: naturally occurring pesticide

compounds containing carbon. These pesticides can have varying toxicity levels and may or may not be appropriate for organic pest control. The connection between organic pest management (control) and organic pesticides is far more limited than the name would imply. *(See also organic pest management)*

Organic Pest Management: Synonymous with organic pest control, however the term organic pest management is more appropriate because it implies the process by which proper treatments are accomplished. Organic pest management is a process with a core set of standards including an intentional effort to limit the use of pesticides, and when pesticides are used, only biorational pesticides are appropriate.

Pesticide: Any compound used to control or mitigate pests. This includes chemicals with a mode of action described as biological control.

Problem areas log: A form left on a property those affected by pests can use to communicate with pest control technicians regarding pest repeated issues, and conversely pest control technicians can communicate their control solution.

Repellency: *(see also aversion)* The natural state of a chemical or substance to drive away insects, or relevant organisms from a specific area. Many pesticides have a repellency level. A repellency level can range from low to high, and with some repellent insecticides, insects react immediately (flushing agents).

Rodenticides: A specialized class of pesticides used for control or mitigation of rodents:

Second Generation Anticoagulant Rodenticides: Fast acting rodenticides not allowed at organic food facilities and should be avoided at any location where organic pest control is being performed.

Spot Treatments: Spot treatments are treatments to a specific area for a particular pest. If a treatment area is larger than 2 square feet, the treatment is not considered a spot treatment.

Surfactant: Non-pesticide compounds used to increase the efficacy of pesticide formulations by lowering the rate at which droplets form. Yucca schidigera extracts are one surfactant commonly used in organic pest control.

Synergists: Compounds used to add to the efficacy of pesticide formulations by increasing the lethal effect of the active ingredients.

Termiticide: Any pesticide compound used for the control or mitigation of termites.

Tolerance (Levels): Tolerance is the pre-defined number of pests allowed on a specific property before actions such as pesticide applications may be warranted. The tolerance level can at times be as low as 1, which is usually the case with health threats such as bed bugs, and ticks. Tolerance levels can also be set by location such as, 1 ant inside and action might be warranted, but the number outside in such a case might be irrelevant. Tolerance levels are set for a specific action such as pesticide applications. *(See also action threshold)*

Trend Log: Log form filled out on a specific time frame regarding pest issues, and what type of trends have been noted on a property. This form is used to verify if control is being properly maintained or to determine new issues otherwise go un-discovered.

USDA: The United States Department of Agriculture is a government agency providing leadership in food, agriculture, and many other areas. The NOP works under the leadership of the USDA, while the OMRI is a non-profit organization accredited by the USDA.

Void Treatments: Treatments done in areas where humans are not likely to contact the treatment area such as the inside of walls.

Yucca Schidigera: A plant species from which many surfactants that add to the efficacy of pesticide formulations are derived.

Index

Action Threshold 9, 11, 13, 19, 20, 51
Ants 76, 77, 78, 79
Aversion 37, 61
Biopesticides 102
Biorational Pesticides 3- 7, 42-44, 89
Boron Based Baits 58, 59, 60
Botanical 24, 51, 52
Carpet Beetles 15, 84, 85, 86
Castor Oil 38, 39
Cedar Oil 54
Cirkil 47,55
Conducive Conditions 12, 17, 31-34
Dermestid Beetles 84
Diatomaceous earth (DE) 61, 98
Dust Mites 84, 86, 87
Fleas 40, 80, 90, 91
Flushing Agents 23
German Cockroaches 92
Global Harmonization 97
Glycol 65,68-69
Gras 46, 47, 64, 69
Green Pest Control Flow Chart 18
Insect Light Traps (ILT's) 15, 16
IPM Plan 16, 19-22, 28, 30, 34, 35, 45
LC50 47
LD50 46,47
Mode of Action 36, 52, 55, 56, 57
Nematodes 40
Neem Oil 47, 54, 55, 98
Non-Pesticidal Repellents 36
NOP 44
No-See-ems 83, 84
Octopamine 53
OMRI 44, 60
Orange Oil 56,57
Organic Pesticides 3,6 ,42, 43, 59
Pantry Moths 74, 75

Pest Monitoring 14
Plant Oils 3, 51-53, 79
Problem Areas Log 26
Pyrethrum 51, 52, 98
Pyrethroids 52, 55, 56, 98
Repellency 22-24, 36-38, 76
Rodenticides 34, 48
Springtails 87, 88, 89
SGAR Rodenticides 34, 35, 48
Surfactants 39, 53, 57, 58
Synergists 52, 57
Tolerance Levels 9-13, 22,42, 79
Trend Log 26, 27
Unknown Bites 80
USDA 44
Wall Void Treatments 70, 71
Yucca Schidigera 39, 58
Five Steps of IPM 11

References

Truman's Scientific Guide to Pest Management Operations
(ISBN#0-929870-64-6)

Ware, George W. The Pesticide Book
(ISBN#0-913702-58-7)

"Easy Air Organic Allergy Relief Spray – Amazing Solutions"
`https://amazing-solutions.com, Available online at time of publishing`, https://amazing-solutions.com/product/easy-air-organic-allergy-relief-spray

"National Organic Program | Agricultural Marketing Service"
`https://www.ams.usda.gov, Available at time of publishing`, https://www.ams.usda.gov/about-ams/programs-offices/national-organic-program

"Natural, Non-Toxic Pesticides and Bug Repellent — Cedarcide"
`www.cedarcide.com, Available at time of publishing`, www.cedarcide.com

"What We Do | Organic Materials Review Institute."
`https://www.omri.org, Available at time of publishing`, https://www.omri.org/what-we-do

"Biorational and Organic Pesticides | UMass Amherst New England Vegetable Guide" `https://nevegetable.org, Available at time of publishing`, https://nevegetable.org/biorational-and-organic-pesticides

"What Is Ipm? / University of California Statewide Integrated Pest Management Program" `http://www2.ipm.ucanr.edu, Available at time of publishing`, http://www2.ipm.ucanr.edu/WhatIsIPM

"Carpet Beetle Dermatitis – Department of Entomology (Penn State)"
`http://ento.psu.edu, available at time of publishing`,
http://ento.psu.edu/extension/factsheets/carpet-beetle-dermatitis

Jacobs, Steven B., Sr. Extension Associate, Dept. of Entomology, "Penn State Entomological Notes Carpet Beetle Dermatitis"
`http://ento.psu.edu/extension/factsheets/pdf/carpetbeetledermatitis`, Published February 2010, Revised February 2015

www.ingramcontent.com/pod-product-compliance
Lightning Source LLC
Chambersburg PA
CBHW070256230526
45470CB00002B/610